PREDICT IT, THEN WIN IT

# SIMPLE 'DISTANCE AND DIFFERENCE' STRATEGY

## TO INCREASE YOUR CHANCES OF WINNING THE LOTTERY BY CHOOSING YOUR OWN NUMBERS OF YOUR CHOICE

SWAMINATHAN

INDIA • SINGAPORE • MALAYSIA

ISBN 979-8-89446-683-5

# CONTENTS

# PREFACE

My dear brothers and sisters, most of you dream of winning the million/billion lottery jackpot.

It is generally believed that the probability of winning the lottery is very low and its random nature does not increase the chance of winning the lottery in any way.

I believe that with your clever mind and a little love for simple math, you can beat the odds and increase your chances of winning the lottery.

I have designed this book around my simple 'Distance and Difference' strategy with simple math and basic arithmetic symbols, focusing on how to use mathematical calculations to increase your chances of winning the lottery.

My book gives you a great start and guides you to choose your perfect winning numbers for each new draw by choosing your own numbers of your choice.

Although I cannot guarantee winning the lottery, I am proud that my book is perfect for finding the right tips and getting the desired results.

In short, my book combines the joy of reading with the acquisition of useful knowledge.

I hope you will appreciate my book and my simple 'Distance and Difference' strategy.

July 23, 2024
CHENNAI-600 071
AUTHOR

# CHAPTER 1

# SIMPLE 'DISTANCE AND DIFFERENCE' STRATEGY

## WITH SIMPLE MATHEMATICS AND BASIC ARITHMETIC SYMBOLS AND ITS APPLICATIONS

Each lottery has its own specific numbers to play with certain rules and regulations.

Although lotteries come in different names, formats, rules and regulations, my simple 'Distance and Difference' strategy is unique and the same for all lotteries.

In particular, this applies to any lottery that allows individuals to select their own numbers of their choice.

If you are interested in choosing your own numbers of your choice, my book will help you.

An example is the internationally popular Powerball lottery in the United States of America. This allows individuals to choose their own numbers of their choice.

Six numbers are required to be selected when purchasing a Powerball lottery ticket. Five numbers from 1 to 69 [in white balls] and one number from 1 to 26 [in red balls]. But you have to face millions of odds to win.

The goal of the game is to match your numbers with the numbers on randomly selected balls during the Powerball drawing.

If all your numbers match the drawn numbers, you win the jackpot.

Your first five numbers do not have to be in the same order as the drawn numbers. But your $6^{th}$ number, the Powerball number must match the number on the red ball.

Even if all of your numbers are not drawn, you still have a chance to win smaller prizes.

## SIMPLE 'DISTANCE AND DIFFERENCE' STRATEGY WITH SIMPLE MATHEMATICS AND BASIC ARITHMETIC SYMBOLS AND ITS APPLICATIONS:

If a single digit number is drawn, add a 0 in front of it to convert it to a two-digit number to find its vibration number with vibration symbol.

To find the vibration number with the vibration symbol for the drawn two-digit number, it is necessary to create a space between the drawn two-digit number. Do this for all the two-digit numbers drawn in the lottery.

After creating the space between each two-digit number drawn, applying the addition and subtraction principle of simple mathematics is the key step and the most important part of my simple 'Distance and Difference' strategy.

Essentially, find the vibration number with the vibration symbol horizontally for each two-digit number drawn. Most importantly, find the vibration numbers with the vibration symbols vertically on the left and right sides of each set of consecutively drawn two-digit numbers.

This step helps predict the vibration numbers of upcoming draw numbers with vibration symbols.

In the following examples, you can easily understand my simple 'Distance and Difference' strategy in a clear way.

Example:

i. The following numbers have been drawn for the January 2, 2020 US Powerball lottery.

49, 53, 57, 59, 62 and 26

The number of the first white ball is 49. If you create a space between 49, it becomes 4….9.

In the same way, you will get the remaining four white ball numbers and one red ball (Powerball) number as follows.

5….3, 5….7, 5….9, 6….2 and 2….6.

ii. The following numbers have been drawn for the January 5, 2020 US Powerball lottery.

01, 11, 21, 25, 54 and 07

After creating a space between each two-digit number, you will get five white ball numbers and one red ball number as follows.

0….1, 1….1, 2….1, 2….5, 5….4 and 0... 7.

So, following are the US Powerball lottery numbers drawn for the above two days (January 2, 2020 & January 5, 2020) after creating a space between each two-digit number.

4…9, 5…3, 5…7, 5…9, 6…2 & 2…6

0…1, 1…1, 2…1, 2…5, 5…4 & 0…7.

Now, using simple math and basic arithmetic symbols, find the vibration numbers for the above numbers with the vibration symbols. This should be done in two parts.

## Part 1

--------

a. Take the first white ball number 4…9 (drawn on January 2, 2020). For 4….9, the horizontal vibration number is 5. Because 4 + 5= 9. Also, the vibration symbol is +. So, the vibration number with vibration symbol for 4….9 is +5.

04

4……9

+5

The sum of two-digit number 49 is shown as 04 over 4....9. (4+9=13, 1+3=4, 4 becomes 04 in two-digit form).

b. Take the first white ball number 0...1 (drawn on January 5, 2020).
For 0....1, the horizontal vibration number is 1. Because 0+1=1. Also, the vibration symbol is +. So, the vibration number with vibration symbol for 0....1 is +1.

01
0....1
+1

The sum of two-digit number 01 is shown as 01 over 0....1. (0+1=1, 1 becomes 01 in two-digit form).

## Part 2

--------

Considering the first white ball numbers 4....9 & 0....1 for the above two days as a set, find the vibration numbers with the vibration symbols perpendicular to the left and right si des.

04
4.......9
+5
-4 -8
01
0.........1
+1

From 4 to 0, the vibration number is 4 in the left-hand vertical direction and its vibration symbol is -. Because, 4-4=0.

Likewise, from 9 to 1, the vibration number is 8 in the right-hand vertical direction and its vibration symbol is -. Because, 9-8=1.

Similarly find vibration numbers with vibration symbols for each set of remaining $2^{nd}$, $3^{rd}$, $4^{th,}$ $5^{th}$ and $6^{th}$ balls.

For your kind reference, in the second chapter of this book, I am providing US Powerball lottery results from January 2, 2020 to August 4, 2022.

Every detail in the second chapter of this book is based on my simple 'Distance and Difference' strategy. Very easy for you to understand and follow.

Choose a specific lottery to follow on a daily basis.

You can start your work on your computer, laptop or a good note book.

Honestly, another part of my strategy is to set aside a certain amount of time each day to work with the lottery. Definitely follow this.

Pay attention to the horizontal vibration number with the vibration symbol of each two-digit number drawn.

Pay close attention to the vibration numbers with the vibration symbols on the left and right sides of each set where the two-digit numbers are drawn.

Calculate the sum of each two-digit number drawn each day and record above the calculated two-digit number. It should also be monitored.

Also, calculate the sum of the six two-digit numbers (5 white ball numbers + 1 red ball number) drawn each day and record it under draw date. It helps predict the sum of the six upcoming winning numbers (Upcoming 5 white ball numbers + 1 red ball number) to some good extent.

If a two-digit number has another number before or after 0, see the following examples to identify the vibration symbol.

Example (1):
Number 04.
Here, '0' precedes 4.
0+4=4. So, the vibration symbol here is +.

Example (2):
Number 20.
Here, '0' follows 2.
2-2=0.
So, the vibration symbol here is -.

Example (3): In a set,

01

+0 +4

05

First find the vibration symbol from 1 to 5 in the right vertical direction. 1+4=5.
Here, the vibration symbol is +.

Generally, if you add 0 with 0 or subtract 0 from 0, the answer is 0. Therefore, + or - both vibration symbols are associated with 0.

Here, the vibration symbol for the left side vibration number '0' should be according to the right side vibration symbol.

Example (4): In a set,

20

-1 -0

10

As shown in example (3) first find the vibration symbol from 2 to 1 in the left vertical direction.

2-1=1.

Here, the vibration symbol is -.

According to example (3), here, the vibration symbol for the right side vibration number '0' should be according to the left side vibration symbol.

Example (5):
In a set,

56
+0 +0
56

(Or)

05
+0 +0
05
+0 +0
05

As mentioned above, for any two-digit number drawn consecutively for 2/3 days or more, the vibration numbers with the vibration symbols on both sides are +0.

Some Important Notes:

I. When comparing with other calculations, carefully note the vibration numbers with the vibration symbols marked vertically on the left and right sides of each set of two-digit numbers drawn in the past. This observation will help you accurately predict upcoming vibration numbers with vibration symbols for accurate upcoming draw numbers corresponding to the last drawn numbers of the lottery,

II. Other calculations mentioned in this book will also help you predict upcoming draw numbers more accurately. Note those calculations as well,

III. After getting my book and choosing the lottery you want to work on, start your work in earnest as described in the book. Also, continue practicing with dedication every day for two to three months. I hope, this exercise will give you a good confidence and make it easier to predict the appropriate winning numbers for each new draw from previously drawn lottery numbers and

IV. Although I cannot guarantee winning the lottery, my book will certainly give you a good start with the guidance you need to succeed.

# CHAPTER 2

# AMERICAN POWERBALL LOTTERY RESULTS

---

**(JANUARY 2, 2020 TO AUGUST 4, 2022 8.30/9.30 IST) IN NEW FORMAT BASED ON MY SIMPLE 'DISTANCE AND DIFFERENCE' STRATEGY WITH SIMPLE MATHEMATICS AND BASIC ARITHMETIC SYMBOLS IN AN EXAMPLE FOR REFERENCE**

| | W/B – 1 | W/B – 2 | W/B – 3 |
|---|---|---|---|
| 2-1-'20 | 04 | 08 | 03 |
| 306 | 4 9 | 5 3 | 5 7 |
| 09 | +5 | -2 | +2 |
| | (-4) (-8) | (-4) (-2) | (-3) (-6) |
| 5-1-'20 | 01 | 02 | 03 |
| 119 | 0 1 | 1 1 | 2 1 |
| 02 | +1 | +0 | -1 |
| | (+0) (+1) | (-1) (+3) | (-2) (+6) |
| 9-1-'20 | 02 | 04 | 07 |
| 134 | 0 2 | 0 4 | 0 7 |
| 08 | +2 | +4 | +7 |
| | (+0) (+1) | (+2) (-3) | (+2) (-4) |
| 12-1-'20 | 03 | 03 | 05 |
| 140 | 0 3 | 2 1 | 2 3 |
| 05 | +3 | -1 | +1 |
| | (+3) (+6) | (+2) (+0) | (+3) (+0) |
| 16-1-'20 | 03 | 05 | 08 |
| 275 | 3 9 | 4 1 | 5 3 |
| 05 | +6 | -3 | -2 |
| | (-1) (-9) | (-2) (+3) | (-2) (+5) |
| 19-1-'20 | 02 | 06 | 02 |
| 224 | 2 0 | 2 4 | 3 8 |
| 08 | -2 | +2 | +5 |
| | (-1) (+1) | (+1) (-1) | (+1)   |

| 23-1-'20 | 02 | 06 | 08 |
|---|---|---|---|
| 222 | 1 1 | 3 3 | 4 4 |
| 06 | +0 | +0 | +0 |
| | (-1) (+1) | (-3) (+6) | (-3) (+3) |

| 26-1-'20 | 02 | 09 | 08 |
|---|---|---|---|
| 149 | 0 2 | 0 9 | 1 7 |
| 05 | +2 | +9 | +6 |
| | (+0) (+7) | (+1) (-7) | (-0) (-2) |

| 30-1-'20 | 09 | 03 | 06 |
|---|---|---|---|
| 129 | 0 9 | 1 2 | 1 5 |
| 03 | +9 | +1 | +4 |

| W/B – 4 | W/B – 5 | R/B | |
|---|---|---|---|
| 05 | 08 | 08 | |
| 5 9 | 6 2 | 2 6 | X2 |
| +4 | -4 | +4 | |
| (-3) (-4) | (-1) (+2) | (-2) (+1) | |
| 07 | 09 | 07 | |
| 2 5 | 5 4 | 0 7 | X2 |
| +3 | -1 | +7 | |
| (+2) (-2) | (+0) (+2) | (+2) (-5) | |
| 07 | 02 | 04 | |
| 4 3 | 5 6 | 2 2 | X4 |
| -1 | +1 | +0 | |
| (-1) (-2) | (+0) (+3) | (-2) (+1) | |

## Simple 'Distance and Difference' Strategy

04 | 05 | 03

3 1 | 5 9 | 0 3 | X2

-2 | +4 | +3

(+2) (+4) | (+1) (-1) | (+1) (+6)

---

01 | 05 | 01

5 5 | 6 8 | 1 9 | X2

+0 | +2 | +8

(+0) (+1) | (+0) (+0) | (-0) (-1)

---

02 | 05 | 09

5 6 | 6 8 | 1 8 | X2

+1 | +2 | +7

(+0) (+3) | (-0) (-1) | (-1) (-0)

---

05 | 04 | 08

5 9 | 6 7 | 0 8 | X3

+4 | +1 | +8

(-2) (-3) | (+0) (+0) | (+1) (+0)

---

09 | 04 | 09

3 6 | 6 7 | 1 8 | X2

+3 | +1 | +7

(-0) (-5) | (-0) (-7) | (-1) (-6)

---

04 | 06 | 02

3 1 | 6 0 | 0 2 | X2

-2 | -6 | +2

| B/F | W/B – 1 | W/B – 2 | W/B – 3 |
|---|---|---|---|
| 30-1-'20 | 09 | 03 | 06 |
| 129 | 0 9 | 1 2 | 1 5 |
| 03 | +9 | +1 | +4 |
| | (+1) (-7) | (+2) (+1) | (+4) (-1) |
| 2-2-'20 | 03 | 06 | 09 |
| 229 | 1 2 | 3 3 | 5 4 |
| 04 | +1 | +0 | -1 |
| | (+1) (+1) | (-0) (-3) | (-2) (+1) |
| 6-2-'20 | 05 | 03 | 08 |
| 188 | 2 3 | 3 0 | 3 5 |
| 08 | +1 | -3 | +2 |
| | (+1) (+2) | (+1) (+9) | (+2) (-5) |
| 9-2-'20 | 08 | 04 | 05 |
| 265 | 3 5 | 4 9 | 5 0 |
| 04 | +2 | +5 | -5 |
| | (-2) (-1) | (-0) (-2) | (+0) (+4) |
| 13-2-'20 | 05 | 02 | 09 |
| 263 | 1 4 | 4 7 | 5 4 |
| 02 | +3 | +3 | -1 |
| | (+0) (+2) | (-1) (-5) | (-2) (+1) |
| 16-2-'20 | 07 | 05 | 08 |
| 168 | 1 6 | 3 2 | 3 5 |
| 06 | +5 | -1 | +2 |
| | (-0) (-6) | (-2) (-0) | (-2) (-0) |

| | | | |
|---|---|---|---|
| 20-2-'20 | 01 | 03 | 06 |
| 131 | 1 0 | 1 2 | 1 5 |
| 05 | -1 | +1 | +4 |
| | (+1) (+5) | (+2) (+5) | (+2) (+4) |

| | | | |
|---|---|---|---|
| 23-2-'20 | 07 | 01 | 03 |
| 235 | 2 5 | 3 7 | 3 9 |
| 01 | +3 | +4 | +6 |
| | (-2) (+3) | (-1) (-0) | (-1) (-0) |

| | | | |
|---|---|---|---|
| 27-2-'20 | 08 | 09 | 02 |
| 171 | 0 8 | 2 7 | 2 9 |
| 09 | +8 | +5 | +7 |

| W/B – 4 | W/B – 5 | R/B | |
|---|---|---|---|
| 04 | 06 | 02 | |
| 3 1 | 6 0 | 0 2 | X2 |
| -2 | -6 | +2 | |
| (+2) (+6) | (+0) (+0) | (+1) (+1) | |
| 03 | 06 | 04 | |
| 5 7 | 6 0 | 1 3 | X4 |
| +2 | -6 | +2 | |
| (-1) (-6) | (-1) (+7) | (-1) (-1) | |
| 05 | 03 | 02 | |
| 4 1 | 5 7 | 0 2 | X3 |
| -3 | +2 | +2 | |
| (+1) (+8) | (+1) (-1) | (+0) (+4) | |

| 05 | 03 | 06 | |
|---|---|---|---|
| 5 9 | 6 6 | 0 6 | X2 |
| +4 | +0 | +6 | |
| (-0) (-4) | (+0) (+2) | (+2) (-1) | |

| 01 | 05 | 07 | |
|---|---|---|---|
| 5 5 | 6 8 | 2 5 | X2 |
| +0 | +2 | +3 | |
| (-2) (+1) | (-2) (-2) | (-2) (-2) | |

| 09 | 01 | 03 | |
|---|---|---|---|
| 3 6 | 4 6 | 0 3 | X3 |
| +3 | +2 | +3 | |
| (-2) (+3) | (+1) (+0) | (+1) (+6) | |

| 01 | 02 | 01 | |
|---|---|---|---|
| 1 9 | 5 6 | 1 9 | X2 |
| +8 | +1 | +8 | |
| (+5) (-8) | (+1) (-4) | (-0) (-8) | |

| 07 | 08 | 02 | |
|---|---|---|---|
| 6 1 | 6 2 | 1 1 | X3 |
| -5 | -4 | +0 | |
| (-3) (+5) | (-2) (+5) | (+1) (+3) | |

| 09 | 02 | 06 | |
|---|---|---|---|
| 3 6 | 4 7 | 2 4 | X3 |
| +3 | +3 | +2 | |

| B/F | W/B – 1 | W/B – 2 | W/B – 3 |
|---|---|---|---|
| 27-2-‘20 | 08 | 09 | 02 |
| 171 | 0 8 | 2 7 | 2 9 |
| 09 | +8 | +5 | +7 |
| | (+2) (-4) | (+2) (-3) | (+2) (-3) |
| 1-3-‘20 | 06 | 08 | 01 |
| 228 | 2 4 | 4 4 | 4 6 |
| 03 | +2 | +0 | +2 |
| | (-1) (+4) | (-0) (-1) | (+1) (+2) |
| 5-3-‘20 | 09 | 07 | 04 |
| 261 | 1 8 | 4 3 | 5 8 |
| 09 | +7 | -1 | +3 |
| | (-1) (-1) | (-3) (+2) | (-3) (-7) |
| 8-3-‘20 | 07 | 06 | 03 |
| 161 | 0 7 | 1 5 | 2 1 |
| 08 | +7 | +4 | -1 |
| | (-0) (-3) | (+1) (+4) | (+2) (+8) |
| 12-3-‘20 | 04 | 02 | 04 |
| 201 | 0 4 | 2 9 | 4 9 |
| 03 | +4 | +7 | +5 |
| | (+0) (+5) | (-0) (-6) | (-2) (-3) |
| 15-3-‘20 | 09 | 05 | 08 |
| 128 | 0 9 | 2 3 | 2 6 |
| 02 | +9 | +1 | +4 |
| | (+1) (-4) | (+0) (+4) | (+2) (-2) |

| 19-3-'20 | 06 | 09 | 08 |
|---|---|---|---|
| 216 | 1 5 | 2 7 | 4 4 |
| 09 | +4 | +5 | +0 |

(-1) (-3) (-0) (-4) (-0) (-4)

| 22-3-'20 | 02 | 05 | 04 |
|---|---|---|---|
| 206 | 0 2 | 2 3 | 4 0 |
| 08 | +2 | +1 | -4 |

(+0) (+3) (-2) (+6) (-2) (+7)

| 26-3-'20 | 05 | 09 | 09 |
|---|---|---|---|
| 138 | 0 5 | 0 9 | 2 7 |
| 03 | +5 | +9 | +5 |

| W/B – 4 | W/B – 5 | R/B | |
|---|---|---|---|
| 09 | 02 | 06 | |
| 3 6 | 4 7 | 2 4 | X3 |
| +3 | +3 | +2 | |

(+2) (-6) (+1) (-6) (-1) (-1)

| 05 | 06 | 04 | |
|---|---|---|---|
| 5 0 | 5 1 | 1 3 | X3 |
| -5 | -4 | +2 | |

(+1) (+0) (+1) (+7) (+0) (+1)

| 06 | 05 | 05 | |
|---|---|---|---|
| 6 0 | 6 8 | 1 4 | X2 |
| -6 | +2 | +3 | |

(-3) (+3) (-0) (-6) (+1) (-1)

| | | | | | | | | | |
|---|---|---|---|---|---|---|---|---|---|
| | 06 | | | 08 | | | 05 | | |
| | 3 3 | | | 6 2 | | | 2 3 | | X2 |
| | +0 | | | -4 | | | +1 | | |
| (+2) | | (-3) | (+0) | | (+5) | (-2) | | (-1) | |
| | 05 | | | 04 | | | 02 | | |
| | 5 0 | | | 6 7 | | | 0 2 | | X4 |
| | -5 | | | +1 | | | +2 | | |
| (-2) | | (-0) | (-3) | | (-5) | (+0) | | (+6) | |
| | 03 | | | 05 | | | 08 | | |
| | 3 0 | | | 3 2 | | | 0 8 | | X3 |
| | -3 | | | -1 | | | +8 | | |
| (+2) | | (+9) | (+3) | | (+1) | (+0) | | (+0) | |
| | 05 | | | 09 | | | 08 | | |
| | 5 9 | | | 6 3 | | | 0 8 | | X4 |
| | +4 | | | -3 | | | +8 | | |
| (+0) | | (+0) | (+0) | | (+6) | (+1) | | (-5) | |
| | 05 | | | 06 | | | 04 | | |
| | 5 9 | | | 6 9 | | | 1 3 | | X2 |
| | +4 | | | +3 | | | +2 | | |
| (-2) | | (-0) | (-2) | | (-7) | (+0) | | (+3) | |
| | 03 | | | 06 | | | 07 | | |
| | 3 9 | | | 4 2 | | | 1 6 | | X2 |
| | +6 | | | +2 | | | +5 | | |

| B/F | | W/B – 1 | | | W/B – 2 | | | W/B – 3 | |
|---|---|---|---|---|---|---|---|---|---|
| 26-3-'20 | | 05 | | | 09 | | | 09 | |
| 138 | | 0 5 | | | 0 9 | | | 2 7 | |
| 03 | | +5 | | | +9 | | | +5 | |
| | (+0) | | (+2) | (+4) | | (-9) | (+2) | | (+1) |
| 29-3-'20 | | 07 | | | 04 | | | 03 | |
| 227 | | 0 7 | | | 4 0 | | | 4 8 | |
| 02 | | +7 | | | -4 | | | +4 | |
| | (+3) | | (-4) | (-1) | | (+5) | (-0) | | (-3) |
| 2-4-'20 | | 06 | | | 08 | | | 09 | |
| 237 | | 3 3 | | | 3 5 | | | 4 5 | |
| 03 | | +0 | | | +2 | | | +1 | |
| | (-3) | | (+5) | (-0) | | (-4) | (-1) | | (+4) |
| 5-4-'20 | | 08 | | | 04 | | | 03 | |
| 165 | | 0 8 | | | 3 1 | | | 3 9 | |
| 03 | | +8 | | | -2 | | | +6 | |
| | (-0) | | (-6) | (+0) | | (+6) | (+0) | | (+0) |
| 9-4-'20 | | 02 | | | 01 | | | 03 | |
| 185 | | 0 2 | | | 3 7 | | | 3 9 | |
| 05 | | +2 | | | +4 | | | +6 | |
| | (+2) | | (+0) | (-1) | | (+2) | (-0) | | (-9) |
| 12-4-'20 | | 04 | | | 02 | | | 03 | |
| 187 | | 2 2 | | | 2 9 | | | 3 0 | |
| 07 | | +0 | | | +7 | | | -3 | |
| | (-1) | | (-2) | (-1) | | (-7) | (+0) | | (+3) |

| 16-4-'20 | 01 | 03 | 06 |
|---|---|---|---|
| 134 | 1 0 | 1 2 | 3 3 |
| 08 | -1 | +1 | +0 |

(-1) (+4) (+3) (+2) (+1) (+3)

| 19-4-'20 | 04 | 08 | 01 |
|---|---|---|---|
| 232 | 0 4 | 4 4 | 4 6 |
| 07 | +4 | +0 | +2 |

(-0) (-3) (-1) (-1) (-1) (-1)

| 23-4-'20 | 01 | 06 | 08 |
|---|---|---|---|
| 202 | 0 1 | 3 3 | 3 5 |
| 04 | +1 | +0 | +2 |

| W/B – 4 | W/B – 5 | R/B | |
|---|---|---|---|
| 03 | 06 | 07 | |
| 3 9 | 4 2 | 1 6 | X2 |
| +6 | -2 | +5 | |

(+2) (-4) (+2) (+4) (-0) (-5)

| 01 | 03 | 02 | |
|---|---|---|---|
| 5 5 | 6 6 | 1 1 | X2 |
| +0 | +0 | +0 | |

(-1) (+3) (-0) (-6) (+0) (+5)

| 03 | 06 | 07 | |
|---|---|---|---|
| 4 8 | 6 0 | 1 6 | X2 |
| +4 | -6 | +5 | |

(-0) (-8) (-2) (+3) (-1) (-2)

04 | 07 | 04
4 0 | 4 3 | 0 4 X3
-4 | -1 | +4
(+0) (+8) | (+1) (+1) | (+0) (+1)

03 | 09 | 05
4 8 | 5 4 | 0 5 X3
+4 | -1 | +5
(-0) (-6) | (-1) (+3) | (+1) (+2)

06 | 02 | 08
4 2 | 4 7 | 1 7 X3
-2 | +3 | +6
(-1) (+4) | (-0) (-6) | (-1) (-5)

09 | 05 | 02
3 6 | 4 1 | 0 2 X3
+3 | -3 | +2
(+2) (+0) | (+2) (+2) | (+1) (+7)

02 | 09 | 01
5 6 | 6 3 | 1 9 X2
+1 | -3 | +8
(-1) (-6) | (+0) (+6) | (+1) (-5)

04 | 06 | 06
4 0 | 6 9 | 2 4 X5
-4 | +3 | +2

| B/F | | W/B – 1 | | | W/B – 2 | | | W/B – 3 | |
|---|---|---|---|---|---|---|---|---|---|
| 23-4-'20 | | 01 | | | 06 | | | 08 | |
| 202 | | 0 1 | | | 3 3 | | | 3 5 | |
| 04 | | +1 | | | +0 | | | +2 | |
| | (+0) | | (+0) | (-3) | | (-0) | (-1) | | (-4) |
| 26-4-'20 | | 01 | | | 03 | | | 03 | |
| 147 | | 0 1 | | | 0 3 | | | 2 1 | |
| 03 | | +1 | | | +3 | | | -1 | |
| | (+0) | | (+1) | (+2) | | (-3) | (+2) | | (+8) |
| 30-4-'20 | | 02 | | | 02 | | | 04 | |
| 219 | | 0 2 | | | 2 0 | | | 4 9 | |
| 03 | | +2 | | | -2 | | | +5 | |
| | (+1) | | (+1) | (-1) | | (+6) | (-1) | | (-6) |
| 3-5-'20 | | 04 | | | 07 | | | 06 | |
| 212 | | 1 3 | | | 1 6 | | | 3 3 | |
| 05 | | +2 | | | +5 | | | +0 | |
| | (-1) | | (+4) | (-1) | | (+2) | (+0) | | (+2) |
| 7-5-'20 | | 07 | | | 08 | | | 08 | |
| 185 | | 0 7 | | | 0 8 | | | 3 5 | |
| 05 | | +7 | | | +8 | | | +2 | |
| | (+1) | | (-5) | (+1) | | (+0) | (+1) | | (-3) |
| 10-5-'20 | | 03 | | | 09 | | | 06 | |
| 204 | | 1 2 | | | 1 8 | | | 4 2 | |
| 06 | | +1 | | | +7 | | | -2 | |
| | (+2) | | (+7) | (+4) | | (-5) | (+1) | | (+2) |

| 14-5-'20 | 03 | 08 | 09 |
|---|---|---|---|
| 279 | 3 9 | 5 3 | 5 4 |
| 09 | +6 | -2 | -1 |
| | (-3) (-1) | (-4) (-1) | (-3) (+2) |

| 17-5-'20 | 08 | 03 | 08 |
|---|---|---|---|
| 138 | 0 8 | 1 2 | 2 6 |
| 03 | +8 | +1 | +4 |
| | (+1) (+0) | (+2) (+2) | (+2) (-6) |

| 21-5-'20 | 09 | 07 | 04 |
|---|---|---|---|
| 193 | 1 8 | 3 4 | 4 0 |
| 04 | +7 | +1 | -4 |

| W/B – 4 | W/B – 5 | R/B | |
|---|---|---|---|
| 04 | 06 | 06 | |
| 4 0 | 6 9 | 2 4 | X5 |
| -4 | +3 | +2 | |
| (+0) (+7) | (-1) (-2) | (-1) (+4) | |

| | | | |
|---|---|---|---|
| 02 | 03 | 09 | |
| 4 7 | 5 7 | 1 8 | X2 |
| +3 | +2 | +7 | |
| (+2) (-6) | (+1) (+0) | (+1) (-8) | |

| | | | |
|---|---|---|---|
| 07 | 04 | 02 | |
| 6 1 | 6 7 | 2 0 | X2 |
| -5 | +1 | -2 | |
| (-1) (+7) | (+0) (+1) | (+0) (+4) | |

04 05 06

5 8 6 8 2 4 X5

+3 +2 +2

(-0) (-8) (-0) (-3) (-0) (-4)

05 02 02

5 0 6 5 2 0 X4

-5 -1 -2

(-1) (+8) (+0) (+0) (-1) (+9)

03 02 01

4 8 6 5 1 9 X5

+4 -1 +8

(+1) (-2) (-1) (+2) (+1) (-9)

02 03 02

5 6 5 7 2 0 X3

+1 +2 -2

(-2) (+3) (-1) (-5) (-1) (+1)

03 06 02

3 9 4 2 1 1 X2

+6 -2 +0

(+1) (-7) (+1) (-2) (-1) (+8)

06 05 09

4 2 5 0 0 9 X2

-2 -5 +9

| B/F | | W/B – 1 | | | W/B – 2 | | | W/B – 3 | |
|---|---|---|---|---|---|---|---|---|---|
| 21-5-'20 | | 09 | | | 07 | | | 04 | |
| 193 | | 1 8 | | | 3 4 | | | 4 0 | |
| 04 | | +7 | | | +1 | | | -4 | |
| | (-1) | | (-6) | (-3) | | (+4) | (-3) | | (+8) |
| 24-5-'20 | | 02 | | | 08 | | | 09 | |
| 88 | | 0 2 | | | 0 8 | | | 1 8 | |
| 07 | | +2 | | | +8 | | | +7 | |
| | (+3) | | (+6) | (+5) | | (+0) | (+4) | | (+1) |
| 28-5-'20 | | 02 | | | 04 | | | 05 | |
| 308 | | 3 8 | | | 5 8 | | | 5 9 | |
| 02 | | +5 | | | +3 | | | +4 | |
| | (-2) | | (-5) | (-2) | | (-6) | (-1) | | (-8) |
| 31-5-'20 | | 04 | | | 05 | | | 05 | |
| 218 | | 1 3 | | | 3 2 | | | 4 1 | |
| 02 | | +2 | | | -1 | | | -3 | |
| | (-1) | | (-2) | (-3) | | (+1) | (-2) | | (+5) |
| 4-6-'20 | | 01 | | | 03 | | | 08 | |
| 152 | | 0 1 | | | 0 3 | | | 2 6 | |
| 08 | | +1 | | | +3 | | | +4 | |
| | (+0) | | (+0) | (+1) | | (+4) | (+1) | | (+2) |
| 7-6-'20 | | 01 | | | 08 | | | 02 | |
| 211 | | 0 1 | | | 1 7 | | | 3 8 | |
| 04 | | +1 | | | +6 | | | +5 | |
| | (+1) | | (-1) | (+2) | | (-4) | (+1) | | (-7) |

| 11-6-'20 | 01 | 06 | 05 |
|---|---|---|---|
| 208 | 1 0 | 3 3 | 4 1 |
| 01 | -1 | +0 | -3 |

(-1) (+2) (-2) (-1) (-1) (+1)

| 14-6-'20 | 02 | 03 | 05 |
|---|---|---|---|
| 166 | 0 2 | 1 2 | 3 2 |
| 04 | +2 | +1 | -1 |

(+0) (+5) (-0) (-2) (+3) (+1)

| 18-6-'20 | 07 | 01 | 09 |
|---|---|---|---|
| 222 | 0 7 | 1 0 | 6 3 |
| 06 | +7 | -1 | -3 |

| W/B – 4 | W/B – 5 | R/B | |
|---|---|---|---|
| 06 | 05 | 09 | |
| 4 2 | 5 0 | 0 9 | X2 |
| -2 | -5 | +9 | |

(-2) (-1) (-3) (+3) (+1) (-3)

| 03 | 05 | 07 | |
|---|---|---|---|
| 2 1 | 2 3 | 1 6 | X4 |
| -1 | +1 | +5 | |

(+4) (+3) (+4) (+5) (+1) (-5)

| 01 | 05 | 03 | |
|---|---|---|---|
| 6 4 | 6 8 | 2 1 | X3 |
| -2 | +2 | -1 | |

(-1) (+4) (-0) (-8) (-1) (+3)

| 04 | 06 | 05 | |
|---|---|---|---|
| 5 8 | 6 0 | 1 4 | X2 |
| +3 | -6 | +3 | |
| (-1) (-7) | (+0) (+4) | (+0) (+3) | |

| 05 | 01 | 08 | |
|---|---|---|---|
| 4 1 | 6 4 | 1 7 | X2 |
| -3 | -2 | +6 | |
| (+2) (+7) | (+0) (+5) | (+0) (+1) | |

| 05 | 06 | 09 | |
|---|---|---|---|
| 6 8 | 6 9 | 1 8 | X2 |
| +2 | +3 | +7 | |
| (-1) (-6) | (-1) (-5) | (+0) (+0) | |

| 07 | 09 | 09 | |
|---|---|---|---|
| 5 2 | 5 4 | 1 8 | X5 |
| -3 | -1 | +7 | |
| (-0) (-2) | (+1) (+1) | (-1) (-3) | |

| 05 | 02 | 05 | |
|---|---|---|---|
| 5 0 | 6 5 | 0 5 | X3 |
| -5 | -1 | +5 | |
| (+1) (+4) | (+0) (+3) | (+1) (-5) | |

| 01 | 05 | 01 | |
|---|---|---|---|
| 6 4 | 6 8 | 1 0 | X3 |
| -2 | +2 | -1 | |

| B/F | W/B – 1 | W/B – 2 | W/B – 3 |
|---|---|---|---|
| 18-6-‘20 | 07 | 01 | 09 |
| 222 | 0 7 | 1 0 | 6 3 |
| 06 | +7 | -1 | -3 |
| | (+1) (-7) | (+2) (+1) | (-2) (-2) |
| 21-6-‘20 | 01 | 04 | 05 |
| 217 | 1 0 | 3 1 | 4 1 |
| 01 | -1 | -2 | -3 |
| | (+0) (+5) | (-1) (+1) | (-2) (+6) |
| 25-6-‘20 | 06 | 04 | 09 |
| 166 | 1 5 | 2 2 | 2 7 |
| 04 | +4 | +0 | +5 |
| | (-1) (+4) | (+1) (+4) | (+2) (+2) |
| 28-6-‘20 | 09 | 09 | 04 |
| 220 | 0 9 | 3 6 | 4 9 |
| 04 | +9 | +3 | +5 |
| | (+1) (-4) | (-1) (+2) | (+1) (-7) |
| 2-7-‘20 | 06 | 01 | 07 |
| 229 | 1 5 | 2 8 | 5 2 |
| 04 | +4 | +6 | -3 |
| | (+0) (+1) | (-0) (-7) | (-3) (+5) |
| 5-7-‘20 | 07 | 03 | 09 |
| 191 | 1 6 | 2 1 | 2 7 |
| 02 | +5 | -1 | +5 |
| | (-1) (-3) | (-1) (-1) | (+1) (-3) |

| 9-7-'20 | 03 | 01 | 07 |
|---|---|---|---|
| 150 | 0 3 | 1 0 | 3 4 |
| 06 | +3 | -1 | +1 |
| | (+1) (+1) | (+0) (+9) | (+3) (-3) |

| 12-7-'20 | 05 | 01 | 07 |
|---|---|---|---|
| 224 | 1 4 | 1 9 | 6 1 |
| 08 | +3 | +8 | -5 |
| | (+1) (+3) | (+3) (-2) | (+0) (+0) |

| 16-7-'20 | 09 | 02 | 07 |
|---|---|---|---|
| 270 | 2 7 | 4 7 | 6 1 |
| 09 | +5 | +3 | -5 |

| W/B – 4 | W/B – 5 | R/B | |
|---|---|---|---|
| 01 | 05 | 01 | |
| 6 4 | 6 8 | 1 0 | X3 |
| -2 | +2 | -1 | |
| (-0) (-1) | (-0) (-1) | (-1) (+5) | |
| 09 | 04 | 05 | |
| 6 3 | 6 7 | 0 5 | X3 |
| -3 | +1 | +5 | |
| (-3) (-0) | (-2) (-1) | (+2) (-2) | |
| 06 | 01 | 05 | |
| 3 3 | 4 6 | 2 3 | X3 |
| +0 | +2 | +1 | |
| (+2) (+3) | (+2) (-4) | (-2) (+5) | |

02
5 6
+1
(-0) (-3)

08
6 2
-4
(+0) (+1)

08
0 8 X2
+8
(+1) (+0)

08
5 3
-2
(+1) (-3)

09
6 3
-3
(-0) (-2)

09
1 8 X4
+7
(-1) (-2)

06
6 0
-6
(-3) (+6)

07
6 1
-5
(+0) (+1)

06
0 6 X2
+6
(-0) (-1)

09
3 6
+3
(+3) (-4)

08
6 2
-4
(+0) (+2)

05
0 5 X10
+5
(-0) (-1)

08
6 2
-4
(+0) (+0)

01
6 4
-2
(+0) (+5)

04
0 4 X2
+4
(+0) (+0)

08
6 2
-4

06
6 9
+3

04
0 4 X10
+4

| B/F | W/B – 1 | W/B – 2 | W/B – 3 |
|---|---|---|---|
| 16-7-‘20 | 09 | 02 | 07 |
| 270 | 2 7 | 4 7 | 6 1 |
| 09 | +5 | +3 | -5 |
| | (-1) (-4) | (-3) (-1) | (-3) (+1) |
| 19-7-‘20 | 04 | 07 | 05 |
| 187 | 1 3 | 1 6 | 3 2 |
| 07 | +2 | +5 | -1 |
| | (+0) (+3) | (+1) (-1) | (+0) (+4) |
| 23-7-‘20 | 07 | 07 | 09 |
| 190 | 1 6 | 2 5 | 3 6 |
| 01 | +5 | +3 | +3 |
| | (-1) (-1) | (-0) (-4) | (+0) (+0) |
| 26-7-‘20 | 05 | 03 | 09 |
| 203 | 0 5 | 2 1 | 3 6 |
| 05 | +5 | -1 | +3 |
| | (+0) (+2) | (+0) (+8) | (-0) (-1) |
| 30-7-‘20 | 07 | 02 | 08 |
| 182 | 0 7 | 2 9 | 3 5 |
| 02 | +7 | +7 | +2 |
| | (-0) (-1) | (-0) (-4) | (+0) (+1) |
| 2-8-‘20 | 06 | 07 | 09 |
| 182 | 0 6 | 2 5 | 3 6 |
| 02 | +6 | +3 | +3 |
| | (+0) (+1) | (-1) (-1) | (-2) (+1) |

| 6-8-'20 | 07 | 05 | 08 |
|---|---|---|---|
| 184 | 0 7 | 1 4 | 1 7 |
| 04 | +7 | +3 | +6 |
| | (-0) (-5) | (-1) (-1) | (-0) (-3) |

| 9-8-'20 | 02 | 03 | 05 |
|---|---|---|---|
| 134 | 0 2 | 0 3 | 1 4 |
| 08 | +2 | +3 | +3 |
| | (+0) (+0) | (+0) (+3) | (+0) (+4) |

| 13-8-'20 | 02 | 06 | 09 |
|---|---|---|---|
| 120 | 0 2 | 0 6 | 1 8 |
| 03 | +2 | +6 | +7 |

| W/B – 4 | W/B – 5 | R/B | |
|---|---|---|---|
| 08 | 06 | 04 | |
| 6 2 | 6 9 | 0 4 | X10 |
| -4 | +3 | +4 | |
| (-1) (+6) | (-1) (-0) | (+0) (+5) | |
| 04 | 05 | 09 | |
| 5 8 | 5 9 | 0 9 | X2 |
| +3 | +4 | +9 | |
| (-1) (-4) | (-0) (-4) | (+1) (-5) | |
| 08 | 01 | 05 | |
| 4 4 | 5 5 | 1 4 | X3 |
| +0 | +0 | +3 | |
| (+2) (-3) | (+1) (-3) | (+0) (+4) | |

| 07 | 08 | 09 | |
|---|---|---|---|
| 6 1 | 6 2 | 1 8 | X2 |
| -5 | -4 | +7 | |
| (-2) (-1) | (-2) (+3) | (+1) (-2) | |

| 04 | 09 | 08 | |
|---|---|---|---|
| 4 0 | 4 5 | 2 6 | X2 |
| -4 | +1 | +4 | |
| (+0) (+3) | (+0) (+3) | (-0) (-2) | |

| 07 | 03 | 06 | |
|---|---|---|---|
| 4 3 | 4 8 | 2 4 | X3 |
| -1 | +4 | +2 | |
| (+1) (+4) | (+2) (-3) | (+0) (+0) | |

| 03 | 02 | 06 | |
|---|---|---|---|
| 5 7 | 6 5 | 2 4 | X5 |
| +2 | -1 | +2 | |
| (-1) (-7) | (-1) (-4) | (+0) (+0) | |

| 04 | 06 | 06 | |
|---|---|---|---|
| 4 0 | 5 1 | 2 4 | X3 |
| -4 | -4 | +2 | |
| (-1) (+6) | (-2) (+6) | (-0) (-3) | |

| 09 | 01 | 03 | |
|---|---|---|---|
| 3 6 | 3 7 | 2 1 | X2 |
| +3 | +4 | -1 | |

| B/F | W/B – 1 | W/B – 2 | W/B – 3 |
|---|---|---|---|
| 13-8-'20 | 02 | 06 | 09 |
| 120 | 0 2 | 0 6 | 1 8 |
| 03 | +2 | +6 | +7 |
| | (+0) (+3) | (+1) (-4) | (+2) (-4) |
| 16-8-'20 | 05 | 03 | 07 |
| 155 | 0 5 | 1 2 | 3 4 |
| 02 | +5 | +1 | +1 |
| | (+1) (-2) | (+1) (+1) | (+1) (+3) |
| 20-8-'20 | 04 | 05 | 02 |
| 219 | 1 3 | 2 3 | 4 7 |
| 03 | +2 | +1 | +3 |
| | (+0) (+6) | (+1) (-3) | (-1) (-1) |
| 23-8-'20 | 01 | 03 | 09 |
| 207 | 1 9 | 3 0 | 3 6 |
| 09 | +8 | -3 | +3 |
| | (-1) (-1) | (-2) (+2) | (-2) (+3) |
| 27-8-'20 | 08 | 03 | 01 |
| 146 | 0 8 | 1 2 | 1 9 |
| 02 | +8 | +1 | +8 |
| | (-0) (-3) | (+1) (-1) | (+1) (-7) |
| 30-8-'20 | 05 | 03 | 04 |
| 130 | 0 5 | 2 1 | 2 2 |
| 04 | +5 | -1 | +0 |
| | (-0) (-4) | (-2) (+3) | (-1) (-1) |

| 3-9-'20 | 01 | 04 | 02 |
|---|---|---|---|
| 123 | 0 1 | 0 4 | 1 1 |
| 06 | +1 | +4 | +0 |
| | (+1) (+4) | (+2) (-3) | (+1) (+1) |

| 6-9-'20 | 06 | 03 | 04 |
|---|---|---|---|
| 139 | 1 5 | 2 1 | 2 2 |
| 04 | +4 | -1 | +0 |
| | (+1) (+2) | (+3) (+1) | (+3) (+3) |

| 10-9-'20 | 09 | 07 | 01 |
|---|---|---|---|
| 279 | 2 7 | 5 2 | 5 5 |
| 09 | +5 | -3 | +0 |

| W/B – 4 | W/B – 5 | R/B | |
|---|---|---|---|
| 09 | 01 | 03 | |
| 3 6 | 3 7 | 2 1 | X2 |
| +3 | +4 | -1 | |
| (+1) (-1) | (+2) (-1) | (-2) (+2) | |
| 09 | 02 | 03 | |
| 4 5 | 5 6 | 0 3 | X3 |
| +1 | +1 | +3 | |
| (+1) (+0) | (+0) (+2) | (+2) (+0) | |
| 01 | 04 | 05 | |
| 5 5 | 5 8 | 2 3 | X10 |
| +0 | +3 | +1 | |
| (-1) (-3) | (+1) (-2) | (-1) (+1) | |

| | | | |
|---|---|---|---|
| 06 | 03 | 05 | |
| 4 2 | 6 6 | 1 4 | X3 |
| -2 | +0 | +3 | |
| (+0) (+5) | (-1) (+2) | (-1) (-2) | |

| | | | |
|---|---|---|---|
| 02 | 04 | 02 | |
| 4 7 | 5 8 | 0 2 | X2 |
| +3 | +3 | +2 | |
| (-2) (+2) | (-1) (-5) | (+1) (-2) | |

| | | | |
|---|---|---|---|
| 02 | 07 | 01 | |
| 2 9 | 4 3 | 1 0 | X2 |
| +7 | -1 | -1 | |
| (-0) (-9) | (+2) (+6) | (+0) (+8) | |

| | | | |
|---|---|---|---|
| 02 | 06 | 09 | |
| 2 0 | 6 9 | 1 8 | X2 |
| -2 | +3 | +7 | |
| (+0) (+7) | (-2) (-2) | (-1) (-1) | |

| | | | |
|---|---|---|---|
| 09 | 02 | 07 | |
| 2 7 | 4 7 | 0 7 | X2 |
| +5 | +3 | +7 | |
| (+4) (-7) | (+2) (-3) | (+2) (-6) | |

| | | | |
|---|---|---|---|
| 06 | 01 | 03 | |
| 6 0 | 6 4 | 2 1 | X3 |
| -6 | -2 | -1 | |

| B/F | W/B – 1 | W/B – 2 | W/B – 3 |
|---|---|---|---|
| 10-9-'20 | 09 | 07 | 01 |
| 279 | 2 7 | 5 2 | 5 5 |
| 09 | +5 | -3 | +0 |
| | (-1) (-1) | (-4) (+5) | (-3) (-5) |
| 13-9-'20 | 07 | 08 | 02 |
| 177 | 1 6 | 1 7 | 2 0 |
| 06 | +5 | +6 | -2 |
| | (-0) (-6) | (+0) (+0) | (+1) (+1) |
| 17-9-'20 | 01 | 08 | 04 |
| 163 | 1 0 | 1 7 | 3 1 |
| 01 | -1 | +6 | -2 |
| | (+0) (+1) | (-0) (-3) | (-1) (+2) |
| 20-9-'20 | 02 | 05 | 05 |
| 166 | 1 1 | 1 4 | 2 3 |
| 04 | +0 | +3 | +1 |
| | (-1) (+7) | (+0) (+3) | (+2) (+6) |
| 24-9-'20 | 08 | 08 | 04 |
| 186 | 0 8 | 1 7 | 4 9 |
| 06 | +8 | +6 | +5 |
| | (+1) (-7) | (+1) (-6) | (-2) (-2) |
| 27-9-'20 | 02 | 03 | 09 |
| 181 | 1 1 | 2 1 | 2 7 |
| 01 | +0 | -1 | +5 |
| | (+0) (+3) | (-1) (+7) | (+1) (-1) |

| 1-10-'20 | 05 | 09 | 09 |
|---|---|---|---|
| 202 | 1 4 | 1 8 | 3 6 |
| 04 | +3 | +7 | +3 |
| | (+0) (+4) | (+2) (-7) | (+0) (+0) |

| 4-10-'20 | 09 | 04 | 09 |
|---|---|---|---|
| 195 | 1 8 | 3 1 | 3 6 |
| 06 | +7 | -2 | +3 |
| | (-1) (-2) | (-1) (+3) | (-0) (-6) |

| 8-10-'20 | 06 | 06 | 03 |
|---|---|---|---|
| 188 | 0 6 | 2 4 | 3 0 |
| 08 | +6 | +2 | -3 |

| W/B – 4 | W/B – 5 | R/B | |
|---|---|---|---|
| 06 | 01 | 03 | |
| 6 0 | 6 4 | 2 1 | X3 |
| -6 | -2 | -1 | |
| (-1) (+3) | (+0) (+3) | (-2) (+3) | |

| 08 | 04 | 04 | |
|---|---|---|---|
| 5 3 | 6 7 | 0 4 | X2 |
| -2 | +1 | +4 | |
| (-0) (-2) | (-1) (-4) | (-0) (-3) | |

| 06 | 08 | 01 | |
|---|---|---|---|
| 5 1 | 5 3 | 0 1 | X2 |
| -4 | -2 | +1 | |
| (-1) (+6) | (+0) (+4) | (+1) (+3) | |

02
4 7
+3
(+1) (-5)

03
5 7
+2
(+0) (+2)

05
1 4 X4
+3
(-1) (-3)

07
5 2
-3
(-2) (+4)

05
5 9
+4
(+1) (-7)

01
0 1 X2
+1
(+2) (+3)

09
3 6
+3
(+1) (+3)

08
6 2
-4
(+0) (+5)

06
2 4 X3
+2
(-1) (+4)

04
4 9
+5
(-0) (-6)

04
6 7
+1
(-2) (-0)

09
1 8 X2
+7
(+1) (-8)

07
4 3
-1
(+1) (+0)

02
4 7
+3
(+1) (-1)

02
2 0 X2
-2
(-1) (+9)

08
5 3
-2

02
5 6
+1

01
1 9 X2
+8

| B/F | W/B – 1 | W/B – 2 | W/B – 3 |
|---|---|---|---|
| 8-10-'20 | 06 | 06 | 03 |
| 188 | 0 6 | 2 4 | 3 0 |
| 08 | +6 | +2 | -3 |
| | (-0) (-1) | (-1) (+4) | (-1) (+3) |
| 11-10-'20 | 05 | 09 | 05 |
| 154 | 0 5 | 1 8 | 2 3 |
| 01 | +5 | +7 | +1 |
| | (+2) (-4) | (+2) (-1) | (+3) (-1) |
| 15-10-'20 | 03 | 01 | 07 |
| 226 | 2 1 | 3 7 | 5 2 |
| 01 | -1 | +4 | -3 |
| | (-2) (+5) | (-2) (-7) | (-2) (-1) |
| 18-10-'20 | 06 | 01 | 04 |
| 151 | 0 6 | 1 0 | 3 1 |
| 07 | +6 | -1 | -2 |
| | (-0) (-5) | (-1) (+3) | (-2) (+2) |
| 22-10-'20 | 01 | 03 | 04 |
| 143 | 0 1 | 0 3 | 1 3 |
| 08 | +1 | +3 | +2 |
| | (+1) (+7) | (+2) (-3) | (+1) (+4) |
| 25-10-'20 | 09 | 02 | 09 |
| 181 | 1 8 | 2 0 | 2 7 |
| 01 | +7 | -2 | +5 |
| | (-0) (-7) | (+0) (+8) | (+1) (+0) |

| 29-10-'20 | 02 | 01 | 01 |
|---|---|---|---|
| 182 | 1 1 | 2 8 | 3 7 |
| 02 | +0 | +6 | +4 |
| | (-1) (+1) | (-2) (-2) | (+1) (-7) |

| 1-11-'20 | 02 | 06 | 04 |
|---|---|---|---|
| 169 | 0 2 | 0 6 | 4 0 |
| 07 | +2 | +6 | -4 |
| | (+2) (+1) | (+3) (-4) | (-1) (+3) |

| 5-11-'20 | 05 | 05 | 06 |
|---|---|---|---|
| 196 | 2 3 | 3 2 | 3 3 |
| 07 | +1 | -1 | +0 |

| W/B – 4 | W/B – 5 | R/B | |
|---|---|---|---|
| 08 | 02 | 01 | |
| 5 3 | 5 6 | 1 9 | X2 |
| -2 | +1 | +8 | |
| (-1) (-3) | (-0) (-6) | (-0) (-1) | |
| 04 | 05 | 09 | |
| 4 0 | 5 0 | 1 8 | X3 |
| -4 | -5 | +7 | |
| (+1) (+3) | (+0) (+8) | (-1) (-3) | |
| 08 | 04 | 05 | |
| 5 3 | 5 8 | 0 5 | X2 |
| -2 | +3 | +5 | |
| (-2) (+4) | (-1) (-4) | (+2) (-2) | |

| | | | |
|---|---|---|---|
| 01 | 08 | 05 | |
| 3 7 | 4 4 | 2 3 | X2 |
| +4 | +0 | +1 | |
| (+1) (-3) | (+1) (+2) | (+0) (+3) | |

| | | | |
|---|---|---|---|
| 08 | 02 | 08 | |
| 4 4 | 5 6 | 2 6 | X3 |
| +0 | +1 | +4 | |
| (+0) (+1) | (+1) (-1) | (-2) (-0) | |

| | | | |
|---|---|---|---|
| 09 | 02 | 06 | |
| 4 5 | 6 5 | 0 6 | X2 |
| +1 | -1 | +6 | |
| (-0) (-5) | (-1) (-2) | (+1) (-3) | |

| | | | |
|---|---|---|---|
| 04 | 08 | 04 | |
| 4 0 | 5 3 | 1 3 | X2 |
| -4 | -2 | +2 | |
| (+0) (+2) | (+0) (+2) | (+1) (+1) | |

| | | | |
|---|---|---|---|
| 06 | 01 | 06 | |
| 4 2 | 5 5 | 2 4 | X3 |
| -2 | +0 | +2 | |
| (+0) (+3) | (-1) (+4) | (-1) (-0) | |

| | | | |
|---|---|---|---|
| 09 | 04 | 05 | |
| 4 5 | 4 9 | 1 4 | X2 |
| +1 | +5 | +3 | |

| B/F | W/B – 1 | W/B – 2 | W/B – 3 |
|---|---|---|---|
| 5-11-'20 | 05 | 05 | 06 |
| 196 | 2 3 | 3 2 | 3 3 |
| 07 | +1 | -1 | +0 |
| | (-1) (+1) | (-2) (+4) | (+0) (+4) |
| 8-11-'20 | 05 | 07 | 01 |
| 191 | 1 4 | 1 6 | 3 7 |
| 02 | +3 | +5 | +4 |
| | (-0) (-1) | (-0) (-1) | (-2) (-0) |
| 12-11-'20 | 04 | 06 | 08 |
| 166 | 1 3 | 1 5 | 1 7 |
| 04 | +2 | +4 | +6 |
| | (-1) (+4) | (+0) (+0) | (+0) (+1) |
| 15-11-'20 | 07 | 06 | 09 |
| 137 | 0 7 | 1 5 | 1 8 |
| 02 | +7 | +4 | +7 |
| | (-0) (-3) | (-1) (-0) | (-0) (-1) |
| 19-11-'20 | 04 | 05 | 08 |
| 126 | 0 4 | 0 5 | 1 7 |
| 09 | +4 | +5 | +6 |
| | (+5) (-3) | (+5) (-1) | (+4) (+0) |
| 22-11-'20 | 06 | 09 | 03 |
| 302 | 5 1 | 5 4 | 5 7 |
| 05 | -4 | -1 | +2 |
| | (-5) (+1) | (+0) (+3) | (+0) (+1) |

| 26-11-'20 | 02 | 03 | 04 |
|---|---|---|---|
| 268 | 0 2 | 5 7 | 5 8 |
| 07 | +2 | +2 | +3 |
| | (+0) (+6) | (-4) (-5) | (-4) (-0) |

| 29-11-'20 | 08 | 03 | 09 |
|---|---|---|---|
| 151 | 0 8 | 1 2 | 1 8 |
| 07 | +8 | +1 | +7 |
| | (+2) (+0) | (+2) (-1) | (+3) (-8) |

| 3-12-'20 | 01 | 04 | 04 |
|---|---|---|---|
| 190 | 2 8 | 3 1 | 4 0 |
| 01 | +6 | -2 | -4 |

| W/B – 4 | W/B – 5 | R/B | |
|---|---|---|---|
| 09 | 04 | 05 | |
| 4 5 | 4 9 | 1 4 | X2 |
| +1 | +5 | +3 | |
| (+0) (+3) | (+1) (-1) | (+0) (+4) | |
| 03 | 04 | 09 | |
| 4 8 | 5 8 | 1 8 | X2 |
| +4 | +3 | +7 | |
| (-0) (-3) | (+1) (-5) | (-0) (-5) | |
| 09 | 09 | 04 | |
| 4 5 | 6 3 | 1 3 | X2 |
| +1 | -3 | +2 | |
| (-1) (-3) | (-2) (+2) | (+1) (-3) | |

| | 05 | | | 09 | | | 02 | | |
|---|---|---|---|---|---|---|---|---|---|
| | 3 2 | | | 4 5 | | | 2 0 | | X2 |
| | -1 | | | +1 | | | -2 | | |
| (+1) | | (+1) | (+1) | | (-3) | (-2) | | (+5) | |
| | 07 | | | 07 | | | 05 | | |
| | 4 3 | | | 5 2 | | | 0 5 | | X2 |
| | -1 | | | -3 | | | +5 | | |
| (+2) | | (-3) | (+1) | | (+7) | (+1) | | (-4) | |
| | 06 | | | 06 | | | 02 | | |
| | 6 0 | | | 6 9 | | | 1 1 | | X2 |
| | -6 | | | +3 | | | +0 | | |
| (+0) | | (+0) | (-0) | | (-4) | (+1) | | (+5) | |
| | 06 | | | 02 | | | 08 | | |
| | 6 0 | | | 6 5 | | | 2 6 | | X2 |
| | -6 | | | -1 | | | +4 | | |
| (-2) | | (+4) | (-1) | | (-4) | (-1) | | (+2) | |
| | 08 | | | 06 | | | 09 | | |
| | 4 4 | | | 5 1 | | | 1 8 | | X2 |
| | +0 | | | -4 | | | +7 | | |
| (-0) | | (-3) | (-1) | | (+5) | (-1) | | (-4) | |
| | 05 | | | 01 | | | 04 | | |
| | 4 1 | | | 4 6 | | | 0 4 | | X3 |
| | -3 | | | +2 | | | +4 | | |

| B/F | W/B – 1 | W/B – 2 | W/B – 3 |
|---|---|---|---|
| 3-12-'20 | 01 | 04 | 04 |
| 190 | 2 8 | 3 1 | 4 0 |
| 01 | +6 | -2 | -4 |
| | (-2) (-5) | (-3) (+3) | (-4) (+6) |
| 6-12-'20 | 03 | 04 | 06 |
| 124 | 0 3 | 0 4 | 0 6 |
| 07 | +3 | +4 | +6 |
| | (+1) (-2) | (+1) (+0) | (+3) (-5) |
| 10-12-'20 | 02 | 05 | 04 |
| 155 | 1 1 | 1 4 | 3 1 |
| 02 | +0 | +3 | -2 |
| | (+0) (+6) | (+4) (+0) | (+2) (+5) |
| 13-12-'20 | 08 | 09 | 02 |
| 279 | 1 7 | 5 4 | 5 6 |
| 09 | +6 | -1 | +1 |
| | (-1) (-3) | (-3) (-1) | (-2) (+1) |
| 17-12-'20 | 04 | 05 | 01 |
| 199 | 0 4 | 2 3 | 3 7 |
| 01 | +4 | +1 | +4 |
| | (+2) (+3) | (+1) (-1) | (-0) (-3) |
| 20-12-'20 | 09 | 05 | 07 |
| 201 | 2 7 | 3 2 | 3 4 |
| 03 | +5 | -1 | +1 |
| | (-2) (-1) | (-2) (+1) | (+0) (+4) |

| 24-12-'20 | 06 | 04 | 02 |
|---|---|---|---|
| 155 | 0 6 | 1 3 | 3 8 |
| 02 | +6 | +2 | +5 |
| | (+1) (-6) | (+1) (+1) | (-1) (-1) |

| 27-12-'20 | 01 | 06 | 09 |
|---|---|---|---|
| 167 | 1 0 | 2 4 | 2 7 |
| 05 | -1 | +2 | +5 |
| | (-1) (+3) | (+2) (-1) | (+2) (-2) |

| 31-12-'20 | 03 | 07 | 09 |
|---|---|---|---|
| 231 | 0 3 | 4 3 | 4 5 |
| 06 | +3 | -1 | +1 |

| W/B – 4 | W/B – 5 | R/B | |
|---|---|---|---|
| 05 | 01 | 04 | |
| 4 1 | 4 6 | 0 4 | X3 |
| -3 | +2 | +4 | |
| (+0) (+7) | (+1) (-3) | (+1) (-4) | |
| 03 | 08 | 01 | |
| 4 8 | 5 3 | 1 0 | X2 |
| +4 | -2 | -1 | |
| (-0) (-1) | (-1) (+5) | (-1) (+4) | |
| 02 | 03 | 04 | |
| 4 7 | 4 8 | 0 4 | X3 |
| +3 | +4 | +4 | |
| (+2) (-4) | (+2) (+1) | (+2) (-4) | |

| | | | |
|---|---|---|---|
| 09 | 06 | 02 | |
| 6 3 | 6 9 | 2 0 | X2 |
| -3 | +3 | -2 | |
| (-0) (-2) | (-0) (-2) | (-2) (+7) | |

| | | | |
|---|---|---|---|
| 07 | 04 | 07 | |
| 6 1 | 6 7 | 0 7 | X2 |
| -5 | +1 | +7 | |
| (-2) (+2) | (-1) (-5) | (+1) (-4) | |

| | | | |
|---|---|---|---|
| 07 | 07 | 04 | |
| 4 3 | 5 2 | 1 3 | X2 |
| -1 | -3 | +2 | |
| (-1) (+6) | (+0) (+1) | (-1) (+3) | |

| | | | |
|---|---|---|---|
| 03 | 08 | 06 | |
| 3 9 | 5 3 | 0 6 | X3 |
| +6 | -2 | +6 | |
| (+0) (-4) | (+0) (+0) | (+1) (+2) | |

| | | | |
|---|---|---|---|
| 08 | 08 | 09 | |
| 3 5 | 5 3 | 1 8 | X2 |
| +2 | -2 | +7 | |
| (+3) (-4) | (+1) (+2) | (-0) (-4) | |

| | | | |
|---|---|---|---|
| 07 | 02 | 05 | |
| 6 1 | 6 5 | 1 4 | X2 |
| -5 | -1 | +3 | |

| B/F | W/B – 1 | W/B – 2 | W/B – 3 |
|---|---|---|---|
| 31-12-'20 | 03 | 07 | 09 |
| 231 | 0 3 | 4 3 | 4 5 |
| 06 | +3 | -1 | +1 |
| | (+0) (+0) | (-4) (+1) | (-3) (-4) |
| 3-1-'21 | 03 | 04 | 02 |
| 131 | 0 3 | 0 4 | 1 1 |
| 05 | +3 | +4 | +0 |
| | (-0) (-2) | (+2) (-4) | (+1) (+1) |
| 7-1-'21 | 01 | 02 | 04 |
| 172 | 0 1 | 2 0 | 2 2 |
| 01 | +1 | -2 | +0 |
| | (+1) (+3) | (+0) (+6) | (+1) (+6) |
| 10-1-'21 | 05 | 08 | 02 |
| 182 | 1 4 | 2 6 | 3 8 |
| 02 | +4 | +4 | +5 |
| | (-1) (-0) | (-1) (+3) | (-1) (-5) |
| 14-1-'21 | 04 | 01 | 05 |
| 134 | 0 4 | 1 9 | 2 3 |
| 08 | +4 | +8 | +1 |
| | (+1) (+0) | (+1) (-9) | (+1) (+6) |
| 17-1-'21 | 05 | 02 | 03 |
| 207 | 1 4 | 2 0 | 3 9 |
| 09 | +3 | -2 | +6 |
| | (+3) (-4) | (+3) (+3) | (+3) (-9) |

21-1-'21
312
06

04
4 0
-4
(-4) (+5)

08
5 3
-2
(-5) (+5)

06
6 0
-6
(-5) (+7)

24-1-'21
99
09

05
0 5
+5
(+1) (+2)

08
0 8
+8
(+3) (-5)

08
1 7
+6
(+2) (-2)

28-1-'21
188
08

08
1 7
+6

06
3 3
+0

08
3 5
+2

W/B – 4
07
6 1
-5
(-2) (-0)

W/B – 5
02
6 5
-1
(+0) (+2)

R/B
05
1 4 X2
+3
(-1) (+1)

05
4 1
-3
(+2) (-1)

04
6 7
+1
(-0) (-1)

05
0 5 X2
+5
(-0) (-2)

06
6 0
-6
(-2) (+5)

03
6 6
+0
(-2) (-0)

03
0 3 X3
+3
(+1) (+0)

| | | | |
|---|---|---|---|
| 09 | 01 | 04 | |
| 4 5 | 4 6 | 1 3 | X2 |
| +1 | +2 | +2 | |
| (-2) (-0) | (+0) (+3) | (+0) (+1) | |
| 07 | 04 | 05 | |
| 2 5 | 4 9 | 1 4 | X2 |
| +3 | +5 | +3 | |
| (+4) (+0) | (+2) (-2) | (-1) (-2) | |
| 02 | 04 | 02 | |
| 6 5 | 6 7 | 0 2 | X3 |
| -1 | +1 | +2 | |
| (+0) (+3) | (+0) (+2) | (+2) (+0) | |
| 05 | 06 | 04 | |
| 6 8 | 6 9 | 2 2 | X3 |
| +2 | +3 | +0 | |
| (-4) (-1) | (-4) (-1) | (-1) (+2) | |
| 09 | 01 | 05 | |
| 2 7 | 2 8 | 1 4 | X3 |
| +5 | +6 | +3 | |
| (+2) (-5) | (+3) (-6) | (-1) (+5) | |
| 06 | 07 | 09 | |
| 4 2 | 5 2 | 0 9 | X3 |
| -2 | -3 | +9 | |

| B/F | | W/B – 1 | | | W/B – 2 | | | W/B – 3 | |
|---|---|---|---|---|---|---|---|---|---|
| 28-1-'21 | | 08 | | | 06 | | | 08 | |
| 188 | | 1 7 | | | 3 3 | | | 3 5 | |
| 08 | | +6 | | | +0 | | | +2 | |
| | -1 | | -6 | -3 | | -1 | -3 | | +2 |
| 31-1-'21 | | 01 | | | 02 | | | 07 | |
| 127 | | 0 1 | | | 0 2 | | | 0 7 | |
| 01 | | +1 | | | +2 | | | +7 | |
| | +0 | | +4 | +3 | | +5 | +4 | | -7 |
| 4-2-'21 | | 05 | | | 01 | | | 04 | |
| 217 | | 0 5 | | | 3 7 | | | 4 0 | |
| 01 | | +5 | | | +4 | | | -4 | |
| | -0 | | -4 | -2 | | -1 | +0 | | +8 |
| 7-2-'21 | | 01 | | | 07 | | | 03 | |
| 187 | | 0 1 | | | 1 6 | | | 4 8 | |
| 07 | | +1 | | | +5 | | | +4 | |
| | +1 | | +4 | +2 | | +3 | +1 | | +0 |
| 11-2-'21 | | 06 | | | 03 | | | 04 | |
| 249 | | 1 5 | | | 3 9 | | | 5 8 | |
| 06 | | +4 | | | +6 | | | +3 | |
| | +1 | | -5 | -1 | | -1 | -2 | | -5 |
| 14-2-'21 | | 02 | | | 01 | | | 06 | |
| 232 | | 2 0 | | | 2 8 | | | 3 3 | |
| 07 | | -2 | | | +6 | | | +0 | |
| | -2 | | +1 | -1 | | -3 | -1 | | -2 |

| 18-2-'21 | 01 | 06 | 03 |
|---|---|---|---|
| 116 | 0 1 | 1 5 | 2 1 |
| 08 | +1 | +4 | -1 |
| | (+0) (+3) | (-1) (+3) | (+0) (+1) |

| 21-2-'21 | 04 | 08 | 04 |
|---|---|---|---|
| 128 | 0 4 | 0 8 | 2 2 |
| 02 | +4 | +8 | +0 |
| | (+0) (+0) | (+3) (-5) | (+2) (+1) |

| 25-2-'21 | 04 | 06 | 07 |
|---|---|---|---|
| 219 | 0 4 | 3 3 | 4 3 |
| 03 | +4 | +0 | -1 |

| W/B – 4 | W/B – 5 | R/B | |
|---|---|---|---|
| 06 | 07 | 09 | |
| 4 2 | 5 2 | 0 9 | X3 |
| -2 | -3 | +9 | |
| (+1) (+0) | (+1) (-1) | (-0) (-5) | |
| 07 | 07 | 04 | |
| 5 2 | 6 1 | 0 4 | X3 |
| -3 | -5 | +4 | |
| (+1) (+2) | (+0) (+5) | (+0) (+1) | |
| 01 | 03 | 05 | |
| 6 4 | 6 6 | 0 5 | X3 |
| -2 | +0 | +5 | |
| (-2) (+5) | (-0) (-1) | (+0) (+3) | |

| | | | | | | |
|---|---|---|---|---|---|---|
| | 04 | | 02 | | 08 | |
| | 4 9 | | 6 5 | | 0 8 | X2 |
| | +5 | | -1 | | +8 | |
| (+2) | | (-6) (+0) | | (+2) (-0) | | (-1) |
| | 09 | | 04 | | 07 | |
| | 6 3 | | 6 7 | | 0 7 | X2 |
| | -3 | | +1 | | +7 | |
| (+0) | | (+0) (+0) | | (+1) (+2) | | (-7) |
| | 09 | | 05 | | 02 | |
| | 6 3 | | 6 8 | | 2 0 | X2 |
| | -3 | | +2 | | -2 | |
| (-3) | | (-1) (-2) | | (-2) (-2) | | (+1) |
| | 05 | | 01 | | 01 | |
| | 3 2 | | 4 6 | | 0 1 | X3 |
| | -1 | | +2 | | +1 | |
| (+0) | | (+0) (+1) | | (+2) (+0) | | (+3) |
| | 05 | | 04 | | 04 | |
| | 3 2 | | 5 8 | | 0 4 | X10 |
| | -1 | | +3 | | +4 | |
| (+2) | | (+1) (+1) | | (-3) (+2) | | (-3) |
| | 08 | | 02 | | 03 | |
| | 5 3 | | 6 5 | | 2 1 | X3 |
| | -2 | | -1 | | -1 | |

| B/F | W/B – 1 | W/B – 2 | W/B – 3 |
| --- | --- | --- | --- |
| 25-2-‘21 | 04 | 06 | 07 |
| 219 | 0 4 | 3 3 | 4 3 |
| 03 | +4 | +0 | -1 |
| | (-0) (-2) | (-1) (+5) | (-1) (-2) |
| 28-2-‘21 | 02 | 01 | 04 |
| 175 | 0 2 | 2 8 | 3 1 |
| 04 | +2 | +6 | -2 |
| | (+2) (-1) | (+2) (-8) | (+1) (+3) |
| 4-3-‘21 | 03 | 04 | 08 |
| 226 | 2 1 | 4 0 | 4 4 |
| 01 | -1 | -4 | +0 |
| | (-1) (-0) | (-1) (+1) | (+1) (-4) |
| 7-3-‘21 | 02 | 04 | 05 |
| 220 | 1 1 | 3 1 | 5 0 |
| 04 | +0 | -2 | -5 |
| | (+0) (+6) | (-2) (+7) | (-2) (+7) |
| 11-3-‘21 | 08 | 09 | 01 |
| 187 | 1 7 | 1 8 | 3 7 |
| 07 | +6 | +7 | +4 |
| | (-1) (-2) | (-0) (-7) | (+2) (-6) |
| 14-3-‘21 | 05 | 02 | 06 |
| 186 | 0 5 | 1 1 | 5 1 |
| 06 | +5 | +0 | -4 |
| | (+3) (-1) | (+2) (+7) | (-1) (+1) |

| 18-3-'21 | 07 | 02 | 06 |
|---|---|---|---|
| 256 | 3 4 | 3 8 | 4 2 |
| 04 | +1 | +5 | -2 |

(-3) (-3) (-3) (-2) (-2) (-0)

| 21-3-'21 | 01 | 06 | 04 |
|---|---|---|---|
| 136 | 0 1 | 0 6 | 2 2 |
| 01 | +1 | +6 | +0 |

(+0) (+3) (+0) (+3) (-1) (+5)

| 25-3-'21 | 04 | 09 | 08 |
|---|---|---|---|
| 113 | 0 4 | 0 9 | 1 7 |
| 05 | +4 | +9 | +6 |

| W/B – 4 | W/B – 5 | R/B | |
|---|---|---|---|
| 08 | 02 | 03 | |
| 5 3 | 6 5 | 2 1 | X3 |
| -2 | -1 | -1 | |

(-1) (+1) (-1) (-3) (-1) (+7)

| 08 | 07 | 09 | |
|---|---|---|---|
| 4 4 | 5 2 | 1 8 | X3 |
| +0 | -3 | +7 | |

(+1) (-4) (+0) (+3) (-0) (-2)

| 05 | 01 | 07 | |
|---|---|---|---|
| 5 0 | 5 5 | 1 6 | X3 |
| -5 | +0 | +5 | |

(+0) (+2) (+0) (+3) (+0) (+2)

| | | | |
|---|---|---|---|
| 07 | 04 | 09 | |
| 5 2 | 5 8 | 1 8 | X4 |
| -3 | +3 | +7 | |
| (-1) (+2) | (-0) (-5) | (+0) (+0) | |
| 08 | 08 | 09 | |
| 4 4 | 5 3 | 1 8 | X3 |
| +0 | -2 | +7 | |
| (+1) (+2) | (+1) (-2) | (-1) (-6) | |
| 02 | 07 | 02 | |
| 5 6 | 6 1 | 0 2 | X2 |
| +1 | -5 | +2 | |
| (+1) (-5) | (+0) (+1) | (+1) (+7) | |
| 07 | 08 | 01 | |
| 6 1 | 6 2 | 1 9 | X2 |
| -5 | -4 | +8 | |
| (-2) (+1) | (-0) (-1) | (-1) (-5) | |
| 06 | 07 | 04 | |
| 4 2 | 6 1 | 0 4 | X3 |
| -2 | -5 | +4 | |
| (-2) (+5) | (-3) (+7) | (+1) (+4) | |
| 09 | 02 | 09 | |
| 2 7 | 3 8 | 1 8 | X2 |
| +5 | +5 | +7 | |

| B/F | W/B – 1 | W/B – 2 | W/B – 3 |
|---|---|---|---|
| 25-3-'21 | 04 | 09 | 08 |
| 113 | 0 4 | 0 9 | 1 7 |
| 05 | +4 | +9 | +6 |
| | (+0) (+2) | (+1) (-5) | (+2) (+1) |
| 28-3-'21 | 06 | 05 | 02 |
| 168 | 0 6 | 1 4 | 3 8 |
| 06 | +6 | +3 | +5 |
| | (-0) (-3) | (-0) (-4) | (+1) (-4) |
| 1-4-'21 | 03 | 01 | 08 |
| 204 | 0 3 | 1 0 | 4 4 |
| 06 | +3 | -1 | +0 |
| | (-0) (-2) | (+0) (+2) | (-3) (+3) |
| 4-4-'21 | 01 | 03 | 08 |
| 127 | 0 1 | 1 2 | 1 7 |
| 01 | +1 | +1 | +6 |
| | (+2) (+6) | (+2) (+3) | (+2) (+2) |
| 8-4-'21 | 09 | 08 | 03 |
| 234 | 2 7 | 3 5 | 3 9 |
| 09 | +5 | +2 | +6 |
| | (-1) (-3) | (-2) (+1) | (-1) (-6) |
| 11-4-'21 | 05 | 07 | 05 |
| 159 | 1 4 | 1 6 | 2 3 |
| 06 | +3 | +5 | +1 |
| | (-0) (-1) | (+2) (-6) | (+1) (+0) |

| 15-4-'21 | 04 | 03 | 06 |
|---|---|---|---|
| 196 | 1 3 | 3 0 | 3 3 |
| 07 | +2 | -3 | +0 |

(-0) (-3) (-1) (+1) (-1) (+3)

| 18-4-'21 | 01 | 03 | 08 |
|---|---|---|---|
| 172 | 1 0 | 2 1 | 2 6 |
| 01 | -1 | -1 | +4 |

(+1) (+1) (+0) (+4) (+1) (-4)

| 22-4-'21 | 03 | 07 | 05 |
|---|---|---|---|
| 214 | 2 1 | 2 5 | 3 2 |
| 07 | -1 | +3 | -1 |

| W/B – 4 | W/B – 5 | R/B | |
|---|---|---|---|
| 09 | 02 | 09 | |
| 2 7 | 3 8 | 1 8 | X2 |
| +5 | +5 | +7 | |

(+1) (+2) (+3) (-3) (-1) (-2)

| 03 | 02 | 06 | |
|---|---|---|---|
| 3 9 | 6 5 | 0 6 | X3 |
| +6 | -1 | +6 | |

(+2) (-4) (+0) (+3) (+2) (-2)

| 01 | 05 | 06 | |
|---|---|---|---|
| 5 5 | 6 8 | 2 4 | X2 |
| +0 | +2 | +2 | |

(-2) (+4) (-1) (-5) (-2) (+1)

| | 03 | | | 08 | | | 05 | | |
|---|---|---|---|---|---|---|---|---|---|
| | 3 9 | | | 5 3 | | | 0 5 | | X2 |
| | +6 | | | -2 | | | +5 | | |
| (+2) | | (-8) | (+1) | | (+3) | (+1) | | (+1) | |

| | 06 | | | 03 | | | 07 | | |
|---|---|---|---|---|---|---|---|---|---|
| | 5 1 | | | 6 6 | | | 1 6 | | X5 |
| | -4 | | | +0 | | | +5 | | |
| (-0) | | (-1) | (-1) | | (-3) | (-1) | | (-3) | |

| | 05 | | | 08 | | | 03 | | |
|---|---|---|---|---|---|---|---|---|---|
| | 5 0 | | | 5 3 | | | 0 3 | | X3 |
| | -5 | | | -2 | | | +3 | | |
| (-1) | | (+5) | (+1) | | (-2) | (+1) | | (+1) | |

| | 09 | | | 07 | | | 05 | | |
|---|---|---|---|---|---|---|---|---|---|
| | 4 5 | | | 6 1 | | | 1 4 | | X2 |
| | +1 | | | -5 | | | +3 | | |
| (-0) | | (-4) | (-2) | | (+8) | (+1) | | (+1) | |

| | 05 | | | 04 | | | 07 | | |
|---|---|---|---|---|---|---|---|---|---|
| | 4 1 | | | 4 9 | | | 2 5 | | X2 |
| | -3 | | | +5 | | | +3 | | |
| (+2) | | (+2) | (+2) | | (-2) | (-2) | | (+1) | |

| | 09 | | | 04 | | | 06 | | |
|---|---|---|---|---|---|---|---|---|---|
| | 6 3 | | | 6 7 | | | 0 6 | | X2 |
| | -3 | | | +1 | | | +6 | | |

| B/F | W/B – 1 | W/B – 2 | W/B – 3 |
|---|---|---|---|
| 22-4-'21<br>214<br>07 | 03<br>2 1<br>-1<br>(+0) (+1) | 07<br>2 5<br>+3<br>(+1) (+1) | 05<br>3 2<br>-1<br>(+1) (+6) |
| 25-4-'21<br>248<br>05 | 04<br>2 2<br>+0<br>(-1) (+4) | 09<br>3 6<br>+3<br>(-2) (+2) | 03<br>4 8<br>+4<br>(-1) (-3) |
| 29-4-'21<br>182<br>02 | 07<br>1 6<br>+5<br>(+2) (-1) | 09<br>1 8<br>+7<br>(+2) (-2) | 08<br>3 5<br>+2<br>(+1) (+2) |
| 2-5-'21<br>245<br>02 | 08<br>3 5<br>+2<br>(-2) (+1) | 09<br>3 6<br>+3<br>(-1) (-3) | 02<br>4 7<br>+3<br>(-2) (+1) |
| 6-5-'21<br>171<br>09 | 07<br>1 6<br>+5<br>(-0) (-4) | 05<br>2 3<br>+1<br>(-1) (+4) | 01<br>2 8<br>+6<br>(-0) (-8) |
| 9-5-'21<br>104<br>05 | 03<br>1 2<br>+1<br>(-1) (-1) | 08<br>1 7<br>+6<br>(+0) (+2) | 02<br>2 0<br>-2<br>(+0) (+0) |

| 13-5-'21 | 01 | 01 | 02 |
|---|---|---|---|
| 149 | 0 1 | 1 9 | 2 0 |
| 05 | +1 | +8 | -2 |
| | (+0) (+3) | (-0) (-9) | (+1) (+7) |

| 16-5-'21 | 04 | 01 | 01 |
|---|---|---|---|
| 183 | 0 4 | 1 0 | 3 7 |
| 03 | +4 | -1 | +4 |
| | (+1) (-3) | (+0) (+3) | (+2) (-2) |

| 20-5-'21 | 02 | 04 | 01 |
|---|---|---|---|
| 208 | 1 1 | 1 3 | 5 5 |
| 01 | +0 | +2 | +0 |

| W/B – 4 | W/B – 5 | R/B | |
|---|---|---|---|
| 09 | 04 | 06 | |
| 6 3 | 6 7 | 0 6 | X2 |
| -3 | +1 | +6 | |
| (-1) (+6) | (-0) (-6) | (+2) (-4) | |

| | | | |
|---|---|---|---|
| 05 | 07 | 04 | |
| 5 9 | 6 1 | 2 2 | X3 |
| +4 | -5 | +0 | |
| (-2) (-0) | (-1) (+2) | (-0) (-1) | |

| | | | |
|---|---|---|---|
| 03 | 08 | 03 | |
| 3 9 | 5 3 | 2 1 | X3 |
| +6 | -2 | -1 | |
| (+3) (-8) | (+1) (+0) | (-2) (+2) | |

| | | | | | | | | | |
|---|---|---|---|---|---|---|---|---|---|
| | 07 | | | 09 | | | 03 | | |
| | 6 1 | | | 6 3 | | | 0 3 | | X3 |
| | -5 | | | -3 | | | +3 | | |
| (-2) | | (-1) | (+0) | | (+0) | (-0) | | (-2) | |

| | | | | | | | | | |
|---|---|---|---|---|---|---|---|---|---|
| | 04 | | | 09 | | | 01 | | |
| | 4 0 | | | 6 3 | | | 0 1 | | X2 |
| | -4 | | | -3 | | | +1 | | |
| (-2) | | (+1) | (-4) | | (+3) | (+0) | | (+7) | |

| | | | | | | | | | |
|---|---|---|---|---|---|---|---|---|---|
| | 03 | | | 08 | | | 08 | | |
| | 2 1 | | | 2 6 | | | 0 8 | | X3 |
| | -1 | | | +4 | | | +8 | | |
| (+1) | | (+7) | (+3) | | (-2) | (+1) | | (-1) | |

| | | | | | | | | | |
|---|---|---|---|---|---|---|---|---|---|
| | 02 | | | 09 | | | 08 | | |
| | 3 8 | | | 5 4 | | | 1 7 | | X2 |
| | +5 | | | -1 | | | +6 | | |
| (+0) | | (+1) | (+1) | | (+5) | (+1) | | (-3) | |

| | | | | | | | | | |
|---|---|---|---|---|---|---|---|---|---|
| | 03 | | | 06 | | | 06 | | |
| | 3 9 | | | 6 9 | | | 2 4 | | X3 |
| | +6 | | | +3 | | | +2 | | |
| (+2) | | (-3) | (+0) | | (+0) | (-2) | | (-0) | |

| | | | | | | | | | |
|---|---|---|---|---|---|---|---|---|---|
| | 02 | | | 06 | | | 04 | | |
| | 5 6 | | | 6 9 | | | 0 4 | | X2 |
| | +1 | | | +3 | | | +4 | | |

| B/F | W/B – 1 | W/B – 2 | W/B – 3 |
|---|---|---|---|
| 20-5-‘21 | 02 | 04 | 01 |
| 208 | 1 1 | 1 3 | 5 5 |
| 01 | +0 | +2 | +0 |
| | (-1) (+2) | (+0) (+6) | (-3) (+2) |
| 23-5-‘21 | 03 | 01 | 09 |
| 134 | 0 3 | 1 9 | 2 7 |
| 08 | +3 | +8 | +5 |
| | (-0) (-1) | (-1) (-1) | (-0) (-6) |
| 27-5-‘21 | 02 | 08 | 03 |
| 143 | 0 2 | 0 8 | 2 1 |
| 08 | +2 | +8 | -1 |
| | (+1) (-1) | (+1) (-5) | (+0) (+1) |
| 30-5-‘21 | 02 | 04 | 04 |
| 139 | 1 1 | 1 3 | 2 2 |
| 04 | +0 | +2 | +0 |
| | (-1) (+5) | (-1) (+4) | (-1) (-1) |
| 3-6-‘21 | 06 | 07 | 02 |
| 176 | 0 6 | 0 7 | 1 1 |
| 05 | +6 | +7 | +0 |
| | (+4) (-2) | (+5) (-5) | (+4) (+3) |
| 6-6-‘21 | 08 | 07 | 09 |
| 309 | 4 4 | 5 2 | 5 4 |
| 03 | +0 | -3 | -1 |
| | (-3) (+5) | (-3) (+6) | (-1) (+2) |

| 10-6-'21 | 01 | 01 | 01 |
|---|---|---|---|
| 206 | 1 9 | 2 8 | 4 6 |
| 08 | +8 | +6 | +2 |
| | (-1) (-1) | (-0) (-3) | (-1) (-2) |

| 13-6-'21 | 08 | 07 | 07 |
|---|---|---|---|
| 156 | 0 8 | 2 5 | 3 4 |
| 03 | +8 | +3 | +1 |
| | (+1) (+1) | (+0) (+4) | (+0) (+0) |

| 17-6-'21 | 01 | 02 | 07 |
|---|---|---|---|
| 201 | 1 9 | 2 9 | 3 4 |
| 03 | +8 | +7 | +1 |

| W/B – 4 | W/B – 5 | R/B | |
|---|---|---|---|
| 02 | 06 | 04 | |
| 5 6 | 6 9 | 0 4 | X2 |
| +1 | +3 | +4 | |
| (-2) (+1) | (-2) (-9) | (+0) (+4) | |
| 01 | 04 | 08 | |
| 3 7 | 4 0 | 0 8 | X2 |
| +4 | -4 | +8 | |
| (-0) (-3) | (+2) (+2) | (+1) (-2) | |
| 07 | 08 | 07 | |
| 3 4 | 6 2 | 1 6 | X2 |
| +1 | -4 | +5 | |
| (-1) (+3) | (-2) (+4) | (+1) (-6) | |

| | | | |
|---|---|---|---|
| 09 | 01 | 02 | |
| 2 7 | 4 6 | 2 0 | X2 |
| +5 | +2 | -2 | |
| (+4) (-1) | (+2) (+1) | (-1) (+9) | |

| | | | |
|---|---|---|---|
| 03 | 04 | 01 | |
| 6 6 | 6 7 | 1 9 | X3 |
| +0 | +1 | +8 | |
| (-0) (-2) | (+0) (+2) | (+1) (-3) | |

| | | | |
|---|---|---|---|
| 01 | 06 | 08 | |
| 6 4 | 6 9 | 2 6 | X3 |
| -2 | +3 | +4 | |
| (-1) (-4) | (-1) (-5) | (-2) (+3) | |

| | | | |
|---|---|---|---|
| 05 | 09 | 09 | |
| 5 0 | 5 4 | 0 9 | X2 |
| -5 | -1 | +9 | |
| (-2) (+8) | (-1) (-3) | (+1) (-9) | |

| | | | |
|---|---|---|---|
| 02 | 05 | 01 | |
| 3 8 | 4 1 | 1 0 | X3 |
| +5 | -3 | -1 | |
| (+1) (-4) | (+1) (-1) | (+1) (+5) | |

| | | | |
|---|---|---|---|
| 08 | 05 | 07 | |
| 4 4 | 5 0 | 2 5 | X2 |
| +0 | -5 | +3 | |

| B/F | W/B – 1 | W/B – 2 | W/B – 3 |
|---|---|---|---|
| 17-6-‘21 | 01 | 02 | 07 |
| 201 | 1 9 | 2 9 | 3 4 |
| 03 | +8 | +7 | +1 |
| | (-1) (-5) | (-0) (-7) | (+0) (+1) |
| 20-6-‘21 | 04 | 04 | 08 |
| 158 | 0 4 | 2 2 | 3 5 |
| 05 | +4 | +0 | +2 |
| | (+1) (-1) | (-0) (-2) | (+1) (-5) |
| 24-6-‘21 | 04 | 02 | 04 |
| 188 | 1 3 | 2 0 | 4 0 |
| 08 | +2 | -2 | -4 |
| | (-1) (+5) | (+1) (+1) | (-1) (+9) |
| 27-6-‘21 | 08 | 04 | 03 |
| 198 | 0 8 | 3 1 | 3 9 |
| 09 | +8 | -2 | +6 |
| | (+2) (-4) | (-1) (+8) | (+2) (-9) |
| 1-7-‘21 | 06 | 02 | 05 |
| 248 | 2 4 | 2 9 | 5 0 |
| 05 | +2 | +7 | -5 |
| | (+0) (+2) | (+2) (-9) | (-1) (+1) |

| 4-7-'21 | 08 | 04 | 05 |
|---|---|---|---|
| 251 | 2 6 | 4 0 | 4 1 |
| 08 | +4 | -4 | -3 |

(-2) (+2) (-2) (+1) (-1) (-1)

| 8-7-'21 | 08 | 03 | 03 |
|---|---|---|---|
| 173 | 0 8 | 2 1 | 3 0 |
| 02 | +8 | -1 | -3 |

(-0) (-7) (-2) (+4) (-1) (+9)

| 11-7-'21 | 01 | 05 | 02 |
|---|---|---|---|
| 154 | 0 1 | 0 5 | 2 9 |
| 01 | +1 | +5 | +7 |

(+3) (+2) (+4) (+1) (+3) (-7)

| 15-7-'21 | 06 | 01 | 07 |
|---|---|---|---|
| 262 | 3 3 | 4 6 | 5 2 |
| 01 | +0 | +2 | -3 |

| W/B – 4 | W/B – 5 | R/B | |
|---|---|---|---|
| 08 | 05 | 07 | |
| 4 4 | 5 0 | 2 5 | X2 |
| +0 | -5 | +3 | |

(-1) (+4) (-2) (+9) (-0) (-5)

| 02 | 03 | 02 | |
|---|---|---|---|
| 3 8 | 3 9 | 2 0 | X2 |
| +5 | +6 | -2 | |

(+2) (-7) (+3) (-6) (-2) (+1)

| | | |
|---|---|---|
| 06 | 09 | 01 |
| 5 1 | 6 3 | 0 1 X3 |
| -4 | -3 | +1 |
| (-1) (+2) | (-0) (-3) | (+1) (+6) |
| 07 | 06 | 08 |
| 4 3 | 6 0 | 1 7 X3 |
| -1 | -6 | +6 |
| (+2) (+2) | (+0) (+6) | (-0) (-3) |
| 02 | 03 | 05 |
| 6 5 | 6 6 | 1 4 X4 |
| -1 | +0 | +3 |
| (-1) (-0) | (-0) (-1) | (+1) (+0) |
| 01 | 02 | 06 |
| 5 5 | 6 5 | 2 4 X2 |
| +0 | -1 | +2 |
| (-1) (+4) | (-1) (+2) | (-2) (+4) |
| 04 | 03 | 08 |
| 4 9 | 5 7 | 0 8 X2 |
| +5 | +2 | +8 |
| (+1) (-5) | (+1) (-5) | (-0) (-5) |
| 09 | 08 | 03 |
| 5 4 | 6 2 | 0 3 X2 |
| -1 | -4 | +3 |
| (+0) (+5) | (+0) (+0) | (+1) (-3) |

| | | | |
|---|---|---|---|
| 05 | 08 | 01 | |
| 5 9 | 6 2 | 1 0 | X2 |
| +4 | -4 | -1 | |

| B/F | W/B – 1 | W/B – 2 | W/B – 3 |
|---|---|---|---|
| 15-7-'21 | 06 | 01 | 07 |
| 262 | 3 3 | 4 6 | 5 2 |
| 01 | +0 | +2 | -3 |
| | (-2) (+2) | (-2) (-4) | (-2) (+6) |
| 18-7-'21 | 06 | 04 | 02 |
| 198 | 1 5 | 2 2 | 3 8 |
| 09 | +4 | +0 | +5 |
| | (+1) (+2) | (+0) (+6) | (+1) (-4) |
| 22-7-'21 | 09 | 01 | 08 |
| 245 | 2 7 | 2 8 | 4 4 |
| 02 | +5 | +6 | +0 |
| | (-2) (-6) | (-2) (-4) | (-3) (-3) |
| 25-7-'21 | 01 | 04 | 02 |
| 152 | 0 1 | 0 4 | 1 1 |
| 08 | +1 | +4 | +0 |
| | (+2) (+4) | (+3) (-4) | (+4) (+2) |
| 29-7-'21 | 07 | 03 | 08 |
| 232 | 2 5 | 3 0 | 5 3 |
| 07 | +3 | -3 | -2 |
| | (-2) (-4) | (-1) (+1) | (-3) (-1) |

| | | | | | | | | | |
|---|---|---|---|---|---|---|---|---|---|
| 1-8-'21 | | 01 | | | 03 | | | 04 | |
| 129 | | 0 1 | | | 2 1 | | | 2 2 | |
| 03 | | +1 | | | -1 | | | +0 | |
| | (+0) | | (+4) | (+0) | | (+0) | (+1) | | (+0) |

| | | | | | | | | | |
|---|---|---|---|---|---|---|---|---|---|
| 5-8-'21 | | 05 | | | 03 | | | 05 | |
| 166 | | 0 5 | | | 2 1 | | | 3 2 | |
| 04 | | +5 | | | -1 | | | -1 | |
| | (+0) | | (+2) | (+0) | | (+3) | (+0) | | (+4) |

| | | | | | | | | | |
|---|---|---|---|---|---|---|---|---|---|
| 8-8-'21 | | 07 | | | 06 | | | 09 | |
| 204 | | 0 7 | | | 2 4 | | | 3 6 | |
| 06 | | +7 | | | +2 | | | +3 | |
| | (+1) | | (-5) | (-1) | | (+4) | (-1) | | (-6) |

| | | | |
|---|---|---|---|
| 12-8-'21 | 03 | 09 | 02 |
| 125 | 1 2 | 1 8 | 2 0 |
| 08 | +1 | +7 | -2 |

| | | | | | | | | | |
|---|---|---|---|---|---|---|---|---|---|
| | W/B – 4 | | | W/B – 5 | | | R/B | | |
| | 05 | | | 08 | | | 01 | | |
| | 5 9 | | | 6 2 | | | 1 0 | | X2 |
| | +4 | | | -4 | | | -1 | | |
| (+0) | | (-5) | (+0) | | (+4) | (-1) | | (+3) | |
| | 09 | | | 03 | | | 03 | | |
| | 5 4 | | | 6 6 | | | 0 3 | | X2 |
| | -1 | | | +0 | | | +3 | | |
| (+1) | | (+3) | (+0) | | (+2) | (+1) | | (-2) | |

| 04 | 05 | 02 | |
|---|---|---|---|
| 6 7 | 6 8 | 1 1 | X2 |
| +1 | +2 | +0 | |
| (-1) (+2) | (-0) (-1) | (-0) (-1) | |

| 05 | 04 | 01 | |
|---|---|---|---|
| 5 9 | 6 7 | 1 0 | X2 |
| +4 | +1 | -1 | |
| (+0) (+0) | (-0) (-7) | (-1) (+5) | |

| 05 | 06 | 05 | |
|---|---|---|---|
| 5 9 | 6 0 | 0 5 | X3 |
| +4 | -6 | +5 | |
| (-2) (-5) | (-2) (+7) | (-0) (-1) | |

| 07 | 02 | 04 | |
|---|---|---|---|
| 3 4 | 4 7 | 0 4 | X2 |
| +1 | +3 | +4 | |
| (+0) (+2) | (+1) (+1) | (+1) (+0) | |

| 09 | 04 | 05 | |
|---|---|---|---|
| 3 6 | 5 8 | 1 4 | X2 |
| +3 | +3 | +3 | |
| (+2) (-2) | (+1) (-8) | (+1) (-1) | |

| 09 | 06 | 05 | |
|---|---|---|---|
| 5 4 | 6 0 | 2 3 | X2 |
| -1 | -6 | +1 | |
| (-3) (+5) | (-3) (-0) | (-1) (+3) | |

| | | | |
|---|---|---|---|
| 02 | 03 | 07 | |
| 2 9 | 3 0 | 1 6 | X3 |
| +7 | -3 | +5 | |

| B/F | | W/B – 1 | | | W/B – 2 | | | W/B – 3 | |
|---|---|---|---|---|---|---|---|---|---|
| 12-8-'21 | | 03 | | | 09 | | | 02 | |
| 125 | | 1 2 | | | 1 8 | | | 2 0 | |
| 08 | | +1 | | | +7 | | | -2 | |
| | (-1) | | (+4) | (+1) | | (-7) | (+2) | | (+9) |
| 15-8-'21 | | 06 | | | 03 | | | 04 | |
| 226 | | 0 6 | | | 2 1 | | | 4 9 | |
| 01 | | +6 | | | -1 | | | +5 | |
| | (+3) | | (-1) | (+1) | | (+5) | (+1) | | (-8) |
| 19-8-'21 | | 08 | | | 09 | | | 06 | |
| 264 | | 3 5 | | | 3 6 | | | 5 1 | |
| 03 | | +2 | | | +3 | | | -4 | |
| | (-2) | | (+1) | (-1) | | (+2) | (-2) | | (+5) |
| 22-8-'21 | | 07 | | | 01 | | | 09 | |
| 182 | | 1 6 | | | 2 8 | | | 3 6 | |
| 02 | | +5 | | | +6 | | | +3 | |
| | (+0) | | (+1) | (+1) | | (-2) | (+1) | | (+1) |
| 24-8-'21 | | 08 | | | 09 | | | 02 | |
| 236 | | 1 7 | | | 3 6 | | | 4 7 | |
| 02 | | +6 | | | +3 | | | +3 | |
| | (+1) | | (+0) | (+0) | | (+3) | (+1) | | (-3) |

| 26-8-'21 | 09 | 03 | 09 |
|---|---|---|---|
| 259 | 2 7 | 3 9 | 5 4 |
| 07 | +5 | +6 | -1 |
| | (-1) (-5) | (-1) (-7) | (-3) (+2) |

| 29-8-'21 | 03 | 04 | 08 |
|---|---|---|---|
| 191 | 1 2 | 2 2 | 2 6 |
| 02 | +1 | +0 | +4 |
| | (-1) (+1) | (-1) (+3) | (+2) (-1) |

| 31-8-'21 | 03 | 06 | 09 |
|---|---|---|---|
| 183 | 0 3 | 1 5 | 4 5 |
| 03 | +3 | +4 | +1 |
| | (+1) (-3) | (+1) (-5) | (-2) (+4) |

| 2-9-'21 | 01 | 02 | 02 |
|---|---|---|---|
| 175 | 1 0 | 2 0 | 2 9 |
| 04 | -1 | -2 | +7 |

| W/B – 4 | W/B – 5 | R/B | |
|---|---|---|---|
| 02 | 03 | 07 | |
| 2 9 | 3 0 | 1 6 | X3 |
| +7 | -3 | +5 | |
| (+4) (-4) | (+3) (+7) | (+0) (+2) | |
| 02 | 04 | 09 | |
| 6 5 | 6 7 | 1 8 | X2 |
| -1 | +1 | +7 | |
| (-1) (-0) | (-0) (-6) | (+1) (-2) | |

01 07 08

5 5 6 1 2 6 X2

+0 -5 +4

(-2) (+4) (-1) (+8) (-2) (-2)

03 05 04

3 9 5 9 0 4 X2

+6 +4 +4

(+3) (-9) (+1) (-8) (+1) (+1)

06 07 06

6 0 6 1 1 5 X3

-6 -5 +4

(-1) (+6) (-1) (+8) (+1) (-1)

02 05 06

5 6 5 9 2 4 X3

+1 +4 +2

(-1) (-0) (+0) (+0) (+0) (+2)

01 05 08

4 6 5 9 2 6 X2

+2 +4 +4

(+1) (-5) (+1) (-8) (-2) (+2)

06 07 08

5 1 6 1 0 8 X2

-4 -5 +8

(-1) (+7) (-1) (-0) (+1) (-1)

| | | | |
|---|---|---|---|
| 03 | 06 | 08 | |
| 4 8 | 5 1 | 1 7 | X2 |
| +4 | -4 | +6 | |

| B/F | W/B – 1 | W/B – 2 | W/B – 3 |
|---|---|---|---|
| 2-9-'21 | 01 | 02 | 02 |
| 175 | 1 0 | 2 0 | 2 9 |
| 04 | -1 | -2 | +7 |
| | (+2) (+2) | (+1) (+5) | (+2) (-9) |
| 5-9-'21 | 05 | 08 | 04 |
| 214 | 3 2 | 3 5 | 4 0 |
| 07 | -1 | +2 | -4 |
| | (-2) (-1) | (-1) (-5) | (-2) (+2) |
| 7-9-'21 | 02 | 02 | 04 |
| 164 | 1 1 | 2 0 | 2 2 |
| 02 | +0 | -2 | +0 |
| | (-1) (+8) | (+0) (+2) | (+2) (-1) |
| 9-9-'21 | 09 | 04 | 05 |
| 201 | 0 9 | 2 2 | 4 1 |
| 03 | +9 | +0 | -3 |
| | (+2) (-9) | (+1) (-1) | (-1) (+7) |
| 12-9-'21 | 02 | 04 | 02 |
| 199 | 2 0 | 3 1 | 3 8 |
| 01 | -2 | -2 | +5 |
| | (+1) (+7) | (+1) (-1) | (+2) (-8) |

| 14-9-'21 | 01 | 04 | 05 |
|---|---|---|---|
| 272 | 3 7 | 4 0 | 5 0 |
| 02 | +4 | -4 | -5 |
| | (-3) (-6) | (-4) (+4) | (-4) (+8) |

| 16-9-'21 | 01 | 04 | 09 |
|---|---|---|---|
| 156 | 0 1 | 0 4 | 1 8 |
| 03 | +1 | +4 | +7 |
| | (+0) (+4) | (+3) (+2) | (+2) (+1) |

| 19-9-'21 | 05 | 09 | 03 |
|---|---|---|---|
| 193 | 0 5 | 3 6 | 3 9 |
| 04 | +5 | +3 | +6 |
| | (+3) (+2) | (+2) (-5) | (+2) (-5) |

| 21-9-'21 | 01 | 06 | 09 |
|---|---|---|---|
| 279 | 3 7 | 5 1 | 5 4 |
| 09 | +4 | -4 | -1 |

| W/B – 4 | W/B – 5 | R/B | |
|---|---|---|---|
| 03 | 06 | 08 | |
| 4 8 | 5 1 | 1 7 | X2 |
| +4 | -4 | +6 | |
| (+1) (-6) | (+0) (+3) | (-1) (-6) | |
| 07 | 09 | 01 | |
| 5 2 | 5 4 | 0 1 | X5 |
| -3 | -1 | +1 | |
| (-2) (+1) | (+0) (+0) | (+2) (+3) | |

| | 06 | | | 09 | | | 06 | | |
|---|---|---|---|---|---|---|---|---|---|
| | 3 3 | | | 5 4 | | | 2 4 | | X2 |
| | +0 | | | -1 | | | +2 | | |
| (+1) | | (+4) | (+1) | | (-3) | (-0) | | (-3) | |

| | 02 | | | 07 | | | 03 | | |
|---|---|---|---|---|---|---|---|---|---|
| | 4 7 | | | 6 1 | | | 2 1 | | X2 |
| | +3 | | | -5 | | | -1 | | |
| (-0) | | (-7) | (-2) | | (+8) | (+0) | | (+0) | |

| | 04 | | | 04 | | | 03 | | |
|---|---|---|---|---|---|---|---|---|---|
| | 4 0 | | | 4 9 | | | 2 1 | | X2 |
| | -4 | | | +5 | | | -1 | | |
| (+2) | | (+1) | (+2) | | (-6) | (+0) | | (+0) | |

| | 07 | | | 09 | | | 03 | | |
|---|---|---|---|---|---|---|---|---|---|
| | 6 1 | | | 6 3 | | | 2 1 | | X3 |
| | -5 | | | -3 | | | -1 | | |
| (-2) | | (+5) | (-0) | | (-1) | (+0) | | (+4) | |

| | 01 | | | 08 | | | 07 | | |
|---|---|---|---|---|---|---|---|---|---|
| | 4 6 | | | 6 2 | | | 2 5 | | X3 |
| | +2 | | | -4 | | | +3 | | |
| (-0) | | (-1) | (-1) | | (+5) | (-1) | | (-4) | |

| | 09 | | | 03 | | | 02 | | |
|---|---|---|---|---|---|---|---|---|---|
| | 4 5 | | | 5 7 | | | 1 1 | | X2 |
| | +1 | | | +2 | | | +0 | | |
| (+1) | | (+3) | (+1) | | (-7) | (+0) | | (+8) | |

| | | | |
|---|---|---|---|
| 04 | 06 | 01 | |
| 5 8 | 6 0 | 1 9 | X2 |
| +3 | -6 | +8 | |

| B/F | | W/B – 1 | | | W/B – 2 | | | W/B – 3 | |
|---|---|---|---|---|---|---|---|---|---|
| 21-9-‘21 | | 01 | | | 06 | | | 09 | |
| 279 | | 3 7 | | | 5 1 | | | 5 4 | |
| 09 | | +4 | | | -4 | | | -1 | |
| | (-1) | | (-7) | (-1) | | (-1) | (-1) | | (+3) |
| 23-9-‘21 | | 02 | | | 04 | | | 02 | |
| 230 | | 2 0 | | | 4 0 | | | 4 7 | |
| 05 | | -2 | | | -4 | | | +3 | |
| | (+0) | | (+2) | (-2) | | (+3) | (-1) | | (-0) |
| 26-9-‘21 | | 04 | | | 05 | | | 01 | |
| 226 | | 2 2 | | | 2 3 | | | 3 7 | |
| 01 | | +0 | | | +1 | | | +4 | |
| | (-0) | | (-1) | (-0) | | (-1) | (+0) | | (+2) |
| 28-9-‘21 | | 03 | | | 04 | | | 03 | |
| 198 | | 2 1 | | | 2 2 | | | 3 9 | |
| 09 | | -1 | | | +0 | | | +6 | |
| | (-2) | | (+1) | (-2) | | (+5) | (-2) | | (-8) |
| 30-9-‘21 | | 02 | | | 07 | | | 02 | |
| 80 | | 0 2 | | | 0 7 | | | 1 1 | |
| 08 | | +2 | | | +7 | | | +0 | |
| | (+2) | | (+6) | (+3) | | (+1) | (+3) | | (+1) |

| 3-10-'21 | 01 | 02 | 06 |
|---|---|---|---|
| 208 | 2 8 | 3 8 | 4 2 |
| 01 | +6 | +5 | -2 |
| | (-1) (-6) | (-1) (-6) | (+1) (+2) |

| 5-10-'21 | 03 | 04 | 09 |
|---|---|---|---|
| 238 | 1 2 | 2 2 | 5 4 |
| 04 | +1 | +0 | -1 |
| | (-1) (-1) | (-1) (+5) | (-0) (-2) |

| 7-10-'21 | 01 | 08 | 07 |
|---|---|---|---|
| 193 | 0 1 | 1 7 | 5 2 |
| 04 | +1 | +6 | -3 |
| | (+1) (+1) | (+0) (+0) | (-2) (-2) |

| 10-10-'21 | 03 | 08 | 03 |
|---|---|---|---|
| 171 | 1 2 | 1 7 | 3 0 |
| 09 | +1 | +6 | -3 |

| W/B – 4 | W/B – 5 | R/B | |
|---|---|---|---|
| 04 | 06 | 01 | |
| 5 8 | 6 0 | 1 9 | X2 |
| +3 | -6 | +8 | |
| (-0) (-3) | (+0) (+3) | (-1) (-4) | |
| 01 | 09 | 05 | |
| 5 5 | 6 3 | 0 5 | X3 |
| +0 | -3 | +5 | |
| (+1) (-3) | (+0) (+0) | (+1) (+4) | |

| | | | |
|---|---|---|---|
| 08 | 09 | 01 | |
| 6 2 | 6 3 | 1 9 | X3 |
| -4 | -3 | +8 | |
| (-2) (+2) | (-0) (-3) | (-0) (-7) | |

| | | | |
|---|---|---|---|
| 08 | 06 | 03 | |
| 4 4 | 6 0 | 1 2 | X2 |
| +0 | -6 | +1 | |
| (-3) (+3) | (-3) (+2) | (-0) (-1) | |

| | | | |
|---|---|---|---|
| 08 | 05 | 02 | |
| 1 7 | 3 2 | 1 1 | X3 |
| +6 | -1 | +0 | |
| (+3) (+0) | (+2) (+0) | (-1) (-0) | |

| | | | |
|---|---|---|---|
| 02 | 07 | 01 | |
| 4 7 | 5 2 | 0 1 | X2 |
| +3 | -3 | +1 | |
| (+2) (-1) | (+1) (+7) | (+1) (+4) | |

| | | | |
|---|---|---|---|
| 03 | 06 | 06 | |
| 6 6 | 6 9 | 1 5 | X2 |
| +0 | +3 | +4 | |
| (-1) (+2) | (-0) (-5) | (-1) (-4) | |

| | | | |
|---|---|---|---|
| 04 | 01 | 01 | |
| 5 8 | 6 4 | 0 1 | X10 |
| +3 | -2 | +1 | |
| (-1) (-3) | (-0) (-2) | (+0) (+4) | |

| | | |
|---|---|---|
| 09 | 08 | 05 |
| 4 5 | 6 2 | 0 5 X2 |
| +1 | -4 | +5 |

| B/F | W/B – 1 | W/B – 2 | W/B – 3 |
|---|---|---|---|
| 10-10-‘21 | 03 | 08 | 03 |
| 171 | 1 2 | 1 7 | 3 0 |
| 09 | +1 | +6 | -3 |
| | (-0) (-1) | (+1) (-7) | (+0) (+3) |
| 12-10-‘21 | 02 | 02 | 06 |
| 192 | 1 1 | 2 0 | 3 3 |
| 03 | +0 | -2 | +0 |
| | (+1) (+2) | (+0) (+9) | (+1) (+4) |
| 14-10-‘21 | 05 | 02 | 02 |
| 233 | 2 3 | 2 9 | 4 7 |
| 08 | +1 | +7 | +3 |
| | (+1) (-3) | (+1) (-8) | (-0) (-6) |
| 17-10-‘21 | 03 | 04 | 05 |
| 195 | 3 0 | 3 1 | 4 1 |
| 06 | -3 | -2 | -3 |
| | (+0) (+0) | (+0) (+1) | (+0) (+7) |
| 19-10-‘21 | 03 | 05 | 03 |
| 238 | 3 0 | 3 2 | 4 8 |
| 04 | -3 | -1 | +4 |
| | (-3) (+7) | (-1) (+7) | (-1) (-2) |

| 21-10-'21 | 07 | 02 | 09 |
|---|---|---|---|
| 161 | 0 7 | 2 9 | 3 6 |
| 08 | +7 | +7 | +3 |
| | (+1) (-7) | (+1) (-9) | (+2) (-5) |

| 24-10-'21 | 01 | 03 | 06 |
|---|---|---|---|
| 231 | 1 0 | 3 0 | 5 1 |
| 06 | -1 | -3 | -4 |
| | (+0) (+0) | (-1) (+7) | (-3) (+8) |

| 26-10-'21 | 01 | 09 | 02 |
|---|---|---|---|
| 192 | 1 0 | 2 7 | 2 9 |
| 03 | -1 | +5 | +7 |
| | (-1) (+3) | (-2) (-1) | (-0) (-3) |

| 28-10-'21 | 03 | 06 | 08 |
|---|---|---|---|
| 138 | 0 3 | 0 6 | 2 6 |
| 03 | +3 | +6 | +4 |

| W/B – 4 | W/B – 5 | R/B | |
|---|---|---|---|
| 09 | 08 | 05 | |
| 4 5 | 6 2 | 0 5 | X2 |
| +1 | -4 | +5 | |
| (-1) (+4) | (+0) (+3) | (+2) (-1) | |

| | | | |
|---|---|---|---|
| 03 | 02 | 06 | |
| 3 9 | 6 5 | 2 4 | X3 |
| +6 | -1 | +2 | |
| (+2) (+0) | (-0) (-5) | (-1) (+1) | |

| | | | | | | | | | |
|---|---|---|---|---|---|---|---|---|---|
| | 05 | | | 06 | | | 06 | | |
| | 5 9 | | | 6 0 | | | 1 5 | | X2 |
| | +4 | | | -6 | | | +4 | | |
| -1 | | -7 | -2 | | +8 | -1 | | -2 | |

| | | | | | | | | | |
|---|---|---|---|---|---|---|---|---|---|
| | 06 | | | 03 | | | 03 | | |
| | 4 2 | | | 4 8 | | | 0 3 | | X3 |
| | -2 | | | +4 | | | +3 | | |
| +1 | | +1 | +2 | | -5 | +1 | | -1 | |

| | | | | | | | | | |
|---|---|---|---|---|---|---|---|---|---|
| | 08 | | | 09 | | | 03 | | |
| | 5 3 | | | 6 3 | | | 1 2 | | X2 |
| | -2 | | | -3 | | | +1 | | |
| -1 | | -2 | -2 | | -0 | -1 | | +3 | |

| | | | | | | | | | |
|---|---|---|---|---|---|---|---|---|---|
| | 05 | | | 07 | | | 05 | | |
| | 4 1 | | | 4 3 | | | 0 5 | | X2 |
| | -3 | | | -1 | | | +5 | | |
| +1 | | +6 | +2 | | +0 | +2 | | -5 | |

| | | | | | | | | | |
|---|---|---|---|---|---|---|---|---|---|
| | 03 | | | 09 | | | 02 | | |
| | 5 7 | | | 6 3 | | | 2 0 | | X2 |
| | +2 | | | -3 | | | -2 | | |
| -1 | | -3 | -1 | | +5 | +0 | | +4 | |

| | | | | | | | | | |
|---|---|---|---|---|---|---|---|---|---|
| | 08 | | | 04 | | | 06 | | |
| | 4 4 | | | 5 8 | | | 2 4 | | X2 |
| | +0 | | | +3 | | | +2 | | |
| -1 | | +1 | -0 | | -7 | -1 | | +3 | |

| 08 | 06 | 08 | |
|---|---|---|---|
| 3 5 | 5 1 | 1 7 | X4 |
| +2 | -4 | +6 | |

| B/F | W/B – 1 | W/B – 2 | W/B – 3 |
|---|---|---|---|
| 28-10-'21 | 03 | 06 | 08 |
| 138 | 0 3 | 0 6 | 2 6 |
| 03 | +3 | +6 | +4 |
| | (+0) (+2) | (+2) (-3) | (+0) (+2) |
| 31-10-'21 | 05 | 05 | 01 |
| 174 | 0 5 | 2 3 | 2 8 |
| 03 | +5 | +1 | +6 |
| | (+0) (+4) | (+0) (+2) | (+1) (-4) |
| 2-11-'21 | 09 | 07 | 07 |
| 165 | 0 9 | 2 5 | 3 4 |
| 03 | +9 | +3 | +1 |
| | (-0) (-8) | (-2) (-3) | (-1) (-0) |
| 4-11-'21 | 01 | 02 | 06 |
| 160 | 0 1 | 0 2 | 2 4 |
| 07 | +1 | +2 | +2 |
| | (+0) (+7) | (+3) (-2) | (+2) (+4) |
| 7-11-'21 | 08 | 03 | 03 |
| 216 | 0 8 | 3 0 | 4 8 |
| 09 | +8 | -3 | +4 |
| | (+2) (-7) | (+1) (+6) | (-0) (-1) |

| 9-11-'21 | 03 | 01 | 02 |
|---|---|---|---|
| 241 | 2 1 | 4 6 | 4 7 |
| 07 | -1 | +2 | +3 |
| | (-1) (+8) | (-2) (-1) | (-0) (-4) |

| 11-11-'21 | 01 | 07 | 07 |
|---|---|---|---|
| 195 | 1 9 | 2 5 | 4 3 |
| 06 | +8 | +3 | -1 |
| | (-1) (-1) | (-1) (-0) | (-2) (+3) |

| 14-11-'21 | 08 | 06 | 08 |
|---|---|---|---|
| 138 | 0 8 | 1 5 | 2 6 |
| 03 | +8 | +4 | +4 |
| | (-0) (-3) | (+2) (-4) | (+1) (-2) |

| 16-11-'21 | 05 | 04 | 07 |
|---|---|---|---|
| 197 | 0 5 | 3 1 | 3 4 |
| 08 | +5 | -2 | +1 |

| W/B – 4 | W/B – 5 | R/B | |
|---|---|---|---|
| 08 | 06 | 08 | |
| 3 5 | 5 1 | 1 7 | X4 |
| +2 | -4 | +6 | |
| (+1) (-2) | (+0) (+5) | (+0) (+2) | |

| | | | |
|---|---|---|---|
| 07 | 02 | 01 | |
| 4 3 | 5 6 | 1 9 | X2 |
| -1 | +1 | +8 | |
| (+0) (+1) | (-1) (-1) | (-1) (-1) | |

| | | | | | | | | | |
|---|---|---|---|---|---|---|---|---|---|
| | 08 | | | 09 | | | 08 | | |
| | 4 4 | | | 4 5 | | | 0 8 | | X3 |
| | +0 | | | +1 | | | +8 | | |
| (+1) | | (-4) | (+1) | | (+2) | (+2) | | (-2) | |
| | 05 | | | 03 | | | 08 | | |
| | 5 0 | | | 5 7 | | | 2 6 | | X2 |
| | -5 | | | +2 | | | +4 | | |
| (+0) | | (+7) | (+1) | | (-3) | (-2) | | (+3) | |
| | 03 | | | 01 | | | 09 | | |
| | 5 7 | | | 6 4 | | | 0 9 | | X2 |
| | +2 | | | -2 | | | +9 | | |
| (+0) | | (+0) | (-0) | | (-2) | (-0) | | (-1) | |
| | 03 | | | 08 | | | 08 | | |
| | 5 7 | | | 6 2 | | | 0 8 | | X3 |
| | +2 | | | -4 | | | +8 | | |
| (-1) | | (-1) | (-2) | | (+6) | (+1) | | (-4) | |
| | 01 | | | 03 | | | 05 | | |
| | 4 6 | | | 4 8 | | | 1 4 | | X2 |
| | +2 | | | +4 | | | +3 | | |
| (-1) | | (-1) | (-0) | | (-3) | (-1) | | (+5) | |
| | 08 | | | 09 | | | 09 | | |
| | 3 5 | | | 4 5 | | | 0 9 | | X3 |
| | +2 | | | +1 | | | +9 | | |
| (+2) | | (-4) | (+1) | | (-2) | (+2) | | (-6) | |

| | | | |
|---|---|---|---|
| 06 | 08 | 05 | |
| 5 1 | 5 3 | 2 3 | X2 |
| -4 | -2 | +1 | |

| B/F | | W/B – 1 | | | W/B – 2 | | | W/B – 3 | |
|---|---|---|---|---|---|---|---|---|---|
| 16-11-'21 | | 05 | | | 04 | | | 07 | |
| 197 | | 0 5 | | | 3 1 | | | 3 4 | |
| 08 | | +5 | | | -2 | | | +1 | |
| | (-0) | | (-2) | (-2) | | (+5) | (+1) | | (+4) |
| 18-11-'21 | | 03 | | | 07 | | | 03 | |
| 180 | | 0 3 | | | 1 6 | | | 4 8 | |
| 09 | | +3 | | | +5 | | | +4 | |
| | (+4) | | (-3) | (+3) | | (-3) | (+0) | | (+0) |
| 21-11-'21 | | 04 | | | 07 | | | 03 | |
| 278 | | 4 0 | | | 4 3 | | | 4 8 | |
| 08 | | -4 | | | -1 | | | +4 | |
| | (-4) | | (+7) | (-2) | | (-3) | (-2) | | (+1) |
| 23-11-'21 | | 07 | | | 02 | | | 02 | |
| 183 | | 0 7 | | | 2 0 | | | 2 9 | |
| 03 | | +7 | | | -2 | | | +7 | |
| | (+1) | | (-7) | (-1) | | (+6) | (+1) | | (-7) |
| 25-11-'21 | | 01 | | | 07 | | | 05 | |
| 203 | | 1 0 | | | 1 6 | | | 3 2 | |
| 05 | | -1 | | | +5 | | | -1 | |
| | (-1) | | (+8) | (+2) | | (-4) | (+2) | | (+3) |

| 28-11-'21 | 08 | 05 | 01 |
|---|---|---|---|
| 235 | 0 8 | 3 2 | 5 5 |
| 01 | +8 | -1 | +0 |
| | (+1) (+0) | (-1) (+4) | (-3) (+3) |

| 30-11-'21 | 09 | 08 | 01 |
|---|---|---|---|
| 174 | 1 8 | 2 6 | 2 8 |
| 03 | +7 | +4 | +6 |
| | (-0) (-6) | (-1) (-1) | (+1) (+0) |

| 2-12-'21 | 03 | 06 | 02 |
|---|---|---|---|
| 209 | 1 2 | 1 5 | 3 8 |
| 02 | +1 | +4 | +5 |
| | (-0) (-2) | (+3) (-5) | (+1) (-3) |

| 5-12-'21 | 01 | 04 | 09 |
|---|---|---|---|
| 220 | 1 0 | 4 0 | 4 5 |
| 04 | -1 | -4 | +1 |

| W/B – 4 | W/B – 5 | R/B | |
|---|---|---|---|
| 06 | 08 | 05 | |
| 5 1 | 5 3 | 2 3 | X2 |
| -4 | -2 | +1 | |
| (+0) (+1) | (+1) (-3) | (-2) (-2) | |
| 07 | 06 | 01 | |
| 5 2 | 6 0 | 0 1 | X3 |
| -3 | -6 | +1 | |
| (+0) (+7) | (+0) (+9) | (+1) (+8) | |

| 05 | 06 | 01 | |
|---|---|---|---|
| 5 9 | 6 9 | 1 9 | X3 |
| +4 | +3 | +8 | |
| (-2) (-1) | (-0) (-2) | (+1) (-7) | |

| 02 | 04 | 04 | |
|---|---|---|---|
| 3 8 | 6 7 | 2 2 | X2 |
| +5 | +1 | +0 | |
| (+3) (-5) | (-0) (-2) | (-1) (+5) | |

| 09 | 02 | 08 | |
|---|---|---|---|
| 6 3 | 6 5 | 1 7 | X2 |
| -3 | -1 | +6 | |
| (+0) (+1) | (+0) (+1) | (-0) (-7) | |

| 01 | 03 | 01 | |
|---|---|---|---|
| 6 4 | 6 6 | 1 0 | X2 |
| -2 | +0 | -1 | |
| (-3) (+4) | (-2) (+1) | (+0) (+7) | |

| 02 | 02 | 08 | |
|---|---|---|---|
| 3 8 | 4 7 | 1 7 | X2 |
| +5 | +3 | +6 | |
| (+2) (-1) | (+2) (-4) | (+1) (-3) | |

| 03 | 09 | 06 | |
|---|---|---|---|
| 5 7 | 6 3 | 2 4 | X2 |
| +2 | -3 | +2 | |
| (-0) (-1) | (+0) (+4) | (-2) (-2) | |

| | | | |
|---|---|---|---|
| 02 | 04 | 02 | |
| 5 6 | 6 7 | 0 2 | X2 |
| +1 | +1 | +2 | |

| B/F | W/B – 1 | W/B – 2 | W/B – 3 |
|---|---|---|---|
| 5-12-‘21 | 01 | 04 | 09 |
| 220 | 1 0 | 4 0 | 4 5 |
| 04 | -1 | -4 | +1 |
| | (-1) (+3) | (-2) (+1) | (-1) (+3) |
| | 03 | 03 | 02 |
| 177 | 0 3 | 2 1 | 3 8 |
| 06 | +3 | -1 | +5 |
| | (+0) (+0) | (-2) (+6) | (-0) (-5) |
| 9-12-‘21 | 03 | 07 | 06 |
| 186 | 0 3 | 0 7 | 3 3 |
| 06 | +3 | +7 | +0 |
| | (+0) (+0) | (+2) (-2) | (+1) (+1) |
| 12-12-‘21 | 03 | 07 | 08 |
| 199 | 0 3 | 2 5 | 4 4 |
| 01 | +3 | +3 | +0 |
| | (+1) (-3) | (+1) (-5) | (-1) (+3) |
| 14-12-‘21 | 01 | 03 | 01 |
| 193 | 1 0 | 3 0 | 3 7 |
| 04 | -1 | -3 | +4 |
| | (+0) (+9) | (-1) (-0) | (+1) (-7) |

| 16-12-'21 | 01 | 02 | 04 |
|---|---|---|---|
| 195 | 1 9 | 2 0 | 4 0 |
| 06 | +8 | -2 | -4 |
| | (-1) (-7) | (-2) (+6) | (-2) (+4) |

| 19-12-'21 | 02 | 06 | 06 |
|---|---|---|---|
| 145 | 0 2 | 0 6 | 2 4 |
| 01 | +2 | +6 | +2 |
| | (+0) (+0) | (+1) (-3) | (-0) (-1) |

| 21-12-'21 | 02 | 04 | 05 |
|---|---|---|---|
| 140 | 0 2 | 1 3 | 2 3 |
| 05 | +2 | +2 | +1 |
| | (+0) (+5) | (+0) (+3) | (-1) (+6) |

| 23-12-'21 | 07 | 07 | 01 |
|---|---|---|---|
| 173 | 0 7 | 1 6 | 1 9 |
| 02 | +7 | +5 | +8 |

| W/B – 4 | W/B – 5 | R/B | |
|---|---|---|---|
| 02 | 04 | 02 | |
| 5 6 | 6 7 | 0 2 | X2 |
| +1 | +1 | +2 | |
| (-0) (-6) | (-1) (+2) | (+0) (+4) | |
| 05 | 05 | 06 | |
| 5 0 | 5 9 | 0 6 | X3 |
| -5 | +4 | +6 | |
| (+0) (+0) | (+1) (+0) | (+2) (-2) | |

| | | | | | | |
|---|---|---|---|---|---|---|
| 05 | 5 | 0 | -5 | (+0) | (+3) | |
| 06 | 6 | 9 | +3 | (-0) | (-5) | |
| 06 | 2 | 4 | -1 | (-1) | (-4) | X2 |
| 08 | 5 | 3 | -2 | (+0) | (+0) | |
| 01 | 6 | 4 | -2 | (-1) | (+5) | |
| 01 | 1 | 0 | -1 | (-1) | (+4) | X2 |
| 08 | 5 | 3 | -2 | (-1) | (-1) | |
| 05 | 5 | 9 | +4 | (+0) | (+0) | |
| 04 | 0 | 4 | +4 | (-1) | (-4) | X2 |
| 06 | 4 | 2 | -2 | (+1) | (-1) | |
| 05 | 5 | 9 | +4 | (+1) | (-8) | |
| 06 | 1 | 5 | +4 | (+0) | (-4) | X3 |
| 06 | 5 | 1 | -4 | (-2) | (+3) | |
| 07 | 6 | 1 | -5 | (+0) | (+5) | |
| 01 | 0 | 1 | +1 | (-1) | (+1) | X2 |
| 07 | 3 | 4 | +1 | (+1) | (+4) | |
| 03 | 6 | 6 | +0 | (+0) | (+2) | |
| 02 | 0 | 2 | +2 | (-1) | (+3) | X4 |

| | | | |
|---|---|---|---|
| 03 | 05 | 06 | |
| 4 8 | 6 8 | 1 5 | X2 |
| +4 | +2 | +4 | |

| B/F | W/B – 1 | W/B – 2 | W/B – 3 |
|---|---|---|---|
| 23-12-'21 | 07 | 07 | 01 |
| 173 | 0 7 | 1 6 | 1 9 |
| 02 | +7 | +5 | +8 |
| (+2) | (+0) | (+1) (+3) | (+3) (-4) |
| 26-12-'21 | 09 | 02 | 09 |
| 216 | 2 7 | 2 9 | 4 5 |
| 09 | +5 | +7 | +1 |
| (+1) | (-1) | (+1) (-1) | (+0) (+0) |
| 28-12-'21 | 09 | 02 | 09 |
| 264 | 3 6 | 3 8 | 4 5 |
| 03 | +3 | +5 | +1 |
| (-3) | (-4) | (-3) (-2) | (-4) (+4) |
| 30-12-'21 | 02 | 06 | 09 |
| 100 | 0 2 | 0 6 | 0 9 |
| 01 | +2 | +6 | +9 |
| (+0) | (+4) | (+1) (-4) | (+3) (+0) |
| 2-1-'22 | 06 | 03 | 03 |
| 162 | 0 6 | 1 2 | 3 9 |
| 09 | +6 | +1 | +6 |
| (-0) | (-4) | (+0) (+1) | (-0) (-7) |

| 4-1-'22 | 02 | 04 | 05 |
|---|---|---|---|
| 150 | 0 2 | 1 3 | 3 2 |
| 06 | +2 | +2 | -1 |
| | (+0) (+4) | (+0) (+1) | (-1) (+3) |

| 6-1-'22 | 06 | 05 | 07 |
|---|---|---|---|
| 141 | 0 6 | 1 4 | 2 5 |
| 06 | +6 | +3 | +3 |
| | (+2) (-6) | (+1) (-3) | (+1) (+1) |

| 9-1-'22 | 02 | 03 | 09 |
|---|---|---|---|
| 215 | 2 0 | 2 1 | 3 6 |
| 08 | -2 | -1 | +3 |
| | (-1) (+4) | (-1) (+6) | (-2) (+2) |

| 11-1-'22 | 05 | 08 | 09 |
|---|---|---|---|
| 106 | 1 4 | 1 7 | 1 8 |
| 07 | +3 | +6 | +7 |

| W/B – 4 | W/B – 5 | R/B | |
|---|---|---|---|
| 03 | 05 | 06 | |
| 4 8 | 6 8 | 1 5 | x2 |
| +4 | +2 | +4 | |
| (+1) (-3) | (-1) (-0) | (-1) (-3) | |
| 01 | 04 | 02 | |
| 5 5 | 5 8 | 0 2 | x3 |
| +0 | +3 | +2 | |
| (+1) (-3) | (+1) (-4) | (+1) (+7) | |

| | | | | | | | | |
|---|---|---|---|---|---|---|---|---|
| | 08 | | | 01 | | | 01 | |
| | 6 2 | | | 6 4 | | | 1 9 | x2 |
| | -4 | | | -2 | | | +8 | |
| (-3) | | (+1) | (-3) | | (+5) | (-0) | | (-8) |
| | 06 | | | 03 | | | 02 | |
| | 3 3 | | | 3 9 | | | 1 1 | x2 |
| | +0 | | | +6 | | | +0 | |
| (+1) | | (+5) | (+2) | | (-9) | (-1) | | (+6) |
| | 03 | | | 05 | | | 07 | |
| | 4 8 | | | 5 0 | | | 0 7 | x2 |
| | +4 | | | -5 | | | +7 | |
| (-1) | | (-5) | (-1) | | (+8) | (+2) | | (-5) |
| | 06 | | | 03 | | | 04 | |
| | 3 3 | | | 4 8 | | | 2 2 | X2 |
| | +0 | | | +4 | | | +0 | |
| (+0) | | (+0) | (-0) | | (-2) | (-1) | | (+5) |
| | 06 | | | 01 | | | 08 | |
| | 3 3 | | | 4 6 | | | 1 7 | x2 |
| | +0 | | | +2 | | | +6 | |
| (+3) | | (-3) | (+2) | | (-1) | (-0) | | (-4) |
| | 06 | | | 02 | | | 04 | |
| | 6 0 | | | 6 5 | | | 1 3 | x10 |
| | -6 | | | -1 | | | +2 | |
| (-4) | | (+1) | (-4) | | (+2) | (-1) | | (+6) |

| | | |
|---|---|---|
| 03 | 09 | 09 |
| 2 1 | 2 7 | 0 9 X2 |
| -1 | +5 | +9 |

| B/F | W/B – 1 | W/B – 2 | W/B – 3 |
|---|---|---|---|
| 11-1-'22 | 05 | 08 | 09 |
| 106 | 1 4 | 1 7 | 1 8 |
| 07 | +3 | +6 | +7 |
| | (-0) (-2) | (+1) (-6) | (+1) (-6) |
| 13-1-'22 | 03 | 03 | 04 |
| 142 | 1 2 | 2 1 | 2 2 |
| 07 | +1 | -1 | +0 |
| | (-1) (+1) | (-1) (+7) | (+1) (+5) |
| 16-1-'22 | 03 | 09 | 01 |
| 181 | 0 3 | 1 8 | 3 7 |
| 01 | +3 | +7 | +4 |
| | (+0) (+6) | (+1) (-4) | (-0) (-2) |
| 18-1-'22 | 09 | 06 | 08 |
| 201 | 0 9 | 2 4 | 3 5 |
| 03 | +9 | +2 | +2 |
| | (+1) (-8) | (-1) (+1) | (+1) (-2) |
| 20-1-'22 | 02 | 06 | 07 |
| 195 | 1 1 | 1 5 | 4 3 |
| 06 | +0 | +4 | -1 |
| | (-1) (+7) | (-0) (-1) | (-1) (-0) |

| 23-1-'22 | 08 | 05 | 06 |
|---|---|---|---|
| 175 | 0 8 | 1 4 | 3 3 |
| 04 | +8 | +3 | +0 |
| | (+1) (-7) | (+1) (+5) | (-0) (-3) |

| 25-1-'22 | 02 | 02 | 03 |
|---|---|---|---|
| 186 | 1 1 | 2 9 | 3 0 |
| 06 | +0 | +7 | -3 |
| | (-1) (+3) | (-1) (-8) | (+0) (+8) |

| 27-1-'22 | 04 | 02 | 02 |
|---|---|---|---|
| 187 | 0 4 | 1 1 | 3 8 |
| 07 | +4 | +0 | +5 |
| | (-0) (-2) | (+0) (+4) | (+0) (+0) |

| 30-1-'22 | 02 | 06 | 02 |
|---|---|---|---|
| 185 | 0 2 | 1 5 | 3 8 |
| 05 | +2 | +4 | +5 |

| W/B – 4 | W/B – 5 | R/B | |
|---|---|---|---|
| 03 | 09 | 09 | |
| 2 1 | 2 7 | 0 9 | x2 |
| -1 | +5 | +9 | |
| (+1) (-1) | (+1) (-4) | (+2) (-5) | |
| 03 | 06 | 06 | |
| 3 0 | 3 3 | 2 4 | x4 |
| -3 | +0 | +2 | |
| (+2) (+1) | (+2) (+6) | (-1) (-1) | |

| | | | | | |
|---|---|---|---|---|---|
| 06 | 5 1 | -4 | (-1) | (+5) | |
| 05 | 5 9 | +4 | (+1) | (-4) | |
| 04 | 1 3 | +2 | (+1) | (-1) | x2 |
| 01 | 4 6 | +2 | (+1) | (-1) | |
| 02 | 6 5 | -1 | (-0) | (-4) | |
| 04 | 2 2 | +0 | (-1) | (-2) | x2 |
| 01 | 5 5 | +0 | (-2) | (+1) | |
| 07 | 6 1 | -5 | (+0) | (+6) | |
| 01 | 1 0 | -1 | (+0) | (+7) | x3 |
| 09 | 3 6 | +3 | (+1) | (+1) | |
| 04 | 6 7 | +1 | (-1) | (-4) | |
| 08 | 1 7 | +6 | (-0) | (-1) | X2 |
| 02 | 4 7 | +3 | (+0) | (+2) | |
| 08 | 5 3 | -2 | (+1) | (+6) | |
| 07 | 1 6 | +5 | (+0) | (+0) | x3 |
| 04 | 4 9 | +5 | (+1) | (-5) | |
| 06 | 6 9 | +3 | (-0) | (-4) | |
| 07 | 1 6 | +5 | (-0) | (-5) | x3 |

| | | | |
|---|---|---|---|
| 09 | 02 | 02 | |
| 5 4 | 6 5 | 1 1 | x4 |
| -1 | -1 | +0 | |

| B/F | W/B – 1 | W/B – 2 | W/B – 3 |
|---|---|---|---|
| 30-1-‘22 | 02 | 06 | 02 |
| 185 | 0 2 | 1 5 | 3 8 |
| 05 | +2 | +4 | +5 |
| | (+1) (-2) | (+0) (+0) | (+2) (-7) |
| 1-2-‘22 | 01 | 06 | 06 |
| 220 | 1 0 | 1 5 | 5 1 |
| 04 | -1 | +4 | -4 |
| | (+0) (+8) | (+1) (+4) | (-2) (+2) |
| 3-2-‘22 | 09 | 02 | 06 |
| 220 | 1 8 | 2 9 | 3 3 |
| 04 | +7 | +7 | +0 |
| | (-1) (-3) | (-1) (-3) | (-1) (+4) |
| 6-2-‘22 | 05 | 07 | 09 |
| 172 | 0 5 | 1 6 | 2 7 |
| 01 | +5 | +5 | +5 |
| | (+0) (+0) | (-0) (-1) | (+1) (+1) |
| 8-2-‘22 | 05 | 06 | 02 |
| 180 | 0 5 | 1 5 | 3 8 |
| 09 | +5 | +4 | +5 |
| | (-0) (-3) | (+0) (+2) | (-0) (-5) |

| 10-2-'22 | 02 | 08 | 06 |
|---|---|---|---|
| 192 | 0 2 | 1 7 | 3 3 |
| 03 | +2 | +6 | +0 |
| | (+0) (+6) | (-0) (-7) | (-1) (-2) |

| 13-2-'22 | 08 | 01 | 03 |
|---|---|---|---|
| 149 | 0 8 | 1 0 | 2 1 |
| 05 | +8 | -1 | -1 |
| | (+1) (-2) | (+1) (+5) | (+0) (+6) |

| 15-2-'22 | 07 | 07 | 09 |
|---|---|---|---|
| 189 | 1 6 | 2 5 | 2 7 |
| 09 | +5 | +3 | +5 |
| | (+1) (-4) | (+1) (-5) | (+2) (-7) |

| 17-2-'22 | 04 | 03 | 04 |
|---|---|---|---|
| 198 | 2 2 | 3 0 | 4 0 |
| 09 | +0 | -3 | -4 |

| W/B – 4 | W/B – 5 | R/B | |
|---|---|---|---|
| 09 | 02 | 02 | |
| 5 4 | 6 5 | 1 1 | x4 |
| -1 | -1 | +0 | |
| (+1) (-3) | (+0) (+4) | (+0) (+3) | |
| 07 | 06 | 05 | |
| 6 1 | 6 9 | 1 4 | x3 |
| -5 | +3 | +3 | |
| (+0) (+1) | (-0) (-6) | (+0) (+1) | |

| | | | | | | | | | |
|---|---|---|---|---|---|---|---|---|---|
| | 08 | | | 09 | | | 06 | | |
| | 6 2 | | | 6 3 | | | 1 5 | | x3 |
| | -4 | | | -3 | | | +4 | | |
| (-3) | | (+7) | (-0) | | (-2) | (+1) | | (-1) | |

| | | | | | | | | | |
|---|---|---|---|---|---|---|---|---|---|
| | 03 | | | 07 | | | 06 | | |
| | 3 9 | | | 6 1 | | | 2 4 | | x2 |
| | +6 | | | -5 | | | +2 | | |
| (+1) | | (-2) | (+0) | | (+4) | (-1) | | (-4) | |

| | | | | | | | | | |
|---|---|---|---|---|---|---|---|---|---|
| | 02 | | | 02 | | | 01 | | |
| | 4 7 | | | 6 5 | | | 1 0 | | x3 |
| | +3 | | | -1 | | | -1 | | |
| (+1) | | (-6) | (-0) | | (-2) | (+1) | | (+6) | |

| | | | | | | | | | |
|---|---|---|---|---|---|---|---|---|---|
| | 06 | | | 09 | | | 08 | | |
| | 5 1 | | | 6 3 | | | 2 6 | | x2 |
| | -4 | | | -3 | | | +4 | | |
| (-1) | | (-0) | (-0) | | (-1) | (-2) | | (+1) | |

| | | | | | | | | | |
|---|---|---|---|---|---|---|---|---|---|
| | 05 | | | 08 | | | 07 | | |
| | 4 1 | | | 6 2 | | | 0 7 | | x3 |
| | -3 | | | -4 | | | +7 | | |
| (+0) | | (+8) | (-1) | | (+3) | (+1) | | (+0) | |

| | | | | | | | | | |
|---|---|---|---|---|---|---|---|---|---|
| | 04 | | | 01 | | | 08 | | |
| | 4 9 | | | 5 5 | | | 1 7 | | x3 |
| | +5 | | | +0 | | | +6 | | |
| (-0) | | (-7) | (-1) | | (+3) | (-0) | | (-1) | |

| 06 | 03 | 07 | |
|---|---|---|---|
| 4 2 | 4 8 | 1 6 | x2 |
| -2 | +4 | +5 | |

| B/F | | W/B – 1 | | | W/B – 2 | | | W/B – 3 | |
|---|---|---|---|---|---|---|---|---|---|
| 17-2-'22 | | 04 | | | 03 | | | 04 | |
| 198 | | 2 2 | | | 3 0 | | | 4 0 | |
| 09 | | +0 | | | -3 | | | -4 | |
| | (-2) | | (+1) | (-2) | | (-0) | (-3) | | (+5) |
| 20-2-'22 | | 03 | | | 01 | | | 06 | |
| 114 | | 0 3 | | | 1 0 | | | 1 5 | |
| 06 | | +3 | | | -1 | | | +4 | |
| | (-0) | | (-1) | (+2) | | (+6) | (+2) | | (+2) |
| 22-2-'22 | | 02 | | | 09 | | | 01 | |
| 192 | | 0 2 | | | 3 6 | | | 3 7 | |
| 03 | | +2 | | | +3 | | | +4 | |
| | (+0) | | (+4) | (-2) | | (+1) | (-1) | | (-6) |
| 24-2-'22 | | 06 | | | 08 | | | 03 | |
| 161 | | 0 6 | | | 1 7 | | | 2 1 | |
| 08 | | +6 | | | +6 | | | -1 | |
| | (+1) | | (-1) | (+2) | | (-5) | (+1) | | (+5) |
| 27-2-'22 | | 06 | | | 05 | | | 09 | |
| 214 | | 1 5 | | | 3 2 | | | 3 6 | |
| 07 | | +4 | | | -1 | | | +3 | |
| | (-1) | | (+2) | (-1) | | (-1) | (+0) | | (+3) |

| 1-3-'22 | 07 | 03 | 03 |
|---|---|---|---|
| 188 | 0 7 | 2 1 | 3 9 |
| 08 | +7 | -1 | +6 |
| (+1) | (+2) (+1) | (+6) (+1) | (-1) |

| 3-3-'22 | 01 | 01 | 03 |
|---|---|---|---|
| 240 | 1 9 | 3 7 | 4 8 |
| 06 | +8 | +4 | +4 |
| (-1) | (-1) (-1) | (-4) (-1) | (-1) |

| 6-3-'22 | 08 | 05 | 01 |
|---|---|---|---|
| 196 | 0 8 | 2 3 | 3 7 |
| 07 | +8 | +1 | +4 |
| (+1) | (-8) (+2) | (+0) (+2) | (-2) |

| 8-3-'22 | 01 | 07 | 01 |
|---|---|---|---|
| 236 | 1 0 | 4 3 | 5 5 |
| 02 | -1 | -1 | +0 |

| W/B – 4 | W/B – 5 | R/B | |
|---|---|---|---|
| 06 | 03 | 07 | |
| 4 2 | 4 8 | 1 6 | x2 |
| -2 | +4 | +5 | |
| (-1) (+1) | (-0) (-6) | (-0) (-5) | |
| 06 | 06 | 02 | |
| 3 3 | 4 2 | 1 1 | x2 |
| +0 | -2 | +0 | |
| (+1) (+2) | (+2) (+7) | (-1) (+2) | |

09
4 5
+1
(-1) (-0)

06
6 9
+3
(-0) (-5)

03
0 3 x2
+3
(+1) (+5)

08
3 5
+2
(+1) (+3)

01
6 4
-2
(+0) (+0)

09
1 8 x2
+7
(+0) (+1)

03
4 8
+4
(-0) (-1)

01
6 4
-2
(-1) (+1)

01
1 9 x3
+8
(+0) (+0)

02
4 7
+3
(+2) (-6)

01
5 5
+0
(+1) (-2)

01
1 9 x2
+8
(-0) (-7)

07
6 1
-5
(-1) (+1)

09
6 3
-3
(+0) (+0)

03
1 2 x2
+1
(+1) (+1)

07
5 2
-3
(+0) (+7)

09
6 3
-3
(+0) (+4)

04
1 3 x2
+2
(-1) (-1)

| | | | |
|---|---|---|---|
| 05 | 04 | 02 | |
| 5 9 | 6 7 | 0 2 | x3 |
| +4 | +1 | +2 | |

| B/F | W/B – 1 | W/B – 2 | W/B – 3 |
|---|---|---|---|
| 8-3-'22 | 01 | 07 | 01 |
| 236 | 1 0 | 4 3 | 5 5 |
| 02 | -1 | -1 | +0 |
| | (+0) (+3) | (-2) (-1) | (-2) (-1) |
| 10-3-'22 | 04 | 04 | 07 |
| 197 | 1 3 | 2 2 | 3 4 |
| 08 | +2 | +0 | +1 |
| | (+0) (+6) | (-0) (-2) | (+0) (+3) |
| 13-3-'22 | 01 | 02 | 01 |
| 184 | 1 9 | 2 0 | 3 7 |
| 04 | +8 | -2 | +4 |
| | (+1) (-8) | (+0) (+8) | (-0) (-5) |
| 15-3-'22 | 03 | 01 | 05 |
| 180 | 2 1 | 2 8 | 3 2 |
| 09 | -1 | +6 | -1 |
| | (-2) (+2) | (+0) (+0) | (+0) (+2) |
| 17-3-'22 | 03 | 01 | 07 |
| 175 | 0 3 | 2 8 | 3 4 |
| 04 | +3 | +6 | +1 |
| | (+0) (+5) | (-2) (+1) | (-2) (+4) |

| 20-3-'22 | 08 | 09 | 09 |
|---|---|---|---|
| 141 | 0 8 | 0 9 | 1 8 |
| 06 | +8 | +9 | +7 |

(-0) (-7) (+1) (-4) (+1) (-2)

| 22-3-'22 | 01 | 06 | 08 |
|---|---|---|---|
| 186 | 0 1 | 1 5 | 2 6 |
| 06 | +1 | +4 | +4 |

(+3) (+0) (+2) (-3) (+1) (+1)

| 24-3-'22 | 04 | 05 | 01 |
|---|---|---|---|
| 210 | 3 1 | 3 2 | 3 7 |
| 03 | -2 | -1 | +4 |

(-3) (+1) (-2) (-2) (+2) (-7)

| 27-3-'22 | 02 | 01 | 05 |
|---|---|---|---|
| 188 | 0 2 | 1 0 | 5 0 |
| 08 | +2 | -1 | -5 |

| W/B – 4 | W/B – 5 | R/B | |
|---|---|---|---|
| 05 | 04 | 02 | |
| 5 9 | 6 7 | 0 2 | x3 |
| +4 | +1 | +2 | |

(-0) (-8) (+0) (+0) (+1) (-2)

| 06 | 04 | 01 | |
|---|---|---|---|
| 5 1 | 6 7 | 1 0 | x2 |
| -4 | +1 | -1 | |

(-2) (+8) (-0) (-6) (-1) (+8)

| | | | | | | | | |
|---|---|---|---|---|---|---|---|---|
| | 03 | | | 07 | | | 08 | |
| | 3 9 | | | 6 1 | | | 0 8 | x2 |
| | +6 | | | -5 | | | +8 | |
| (+1) | | (-5) | (-2) | | (+8) | (-0) | | (-2) |

| | | | | | | | | |
|---|---|---|---|---|---|---|---|---|
| | 08 | | | 04 | | | 06 | |
| | 4 4 | | | 4 9 | | | 0 6 | x3 |
| | +0 | | | +5 | | | +6 | |
| (-1) | | (+1) | (+1) | | (-1) | (+1) | | (+1) |

| | | | | | | | | |
|---|---|---|---|---|---|---|---|---|
| | 08 | | | 04 | | | 08 | |
| | 3 5 | | | 5 8 | | | 1 7 | x2 |
| | +2 | | | +3 | | | +6 | |
| (+1) | | (+3) | (-0) | | (-6) | (-1) | | (-1) |

| | | | | | | | | |
|---|---|---|---|---|---|---|---|---|
| | 03 | | | 07 | | | 06 | |
| | 4 8 | | | 5 2 | | | 0 6 | X2 |
| | +4 | | | -3 | | | +6 | |
| (+2) | | (-5) | (+1) | | (+3) | (+1) | | (+0) |

| | | | | | | | | |
|---|---|---|---|---|---|---|---|---|
| | 09 | | | 02 | | | 07 | |
| | 6 3 | | | 6 5 | | | 1 6 | x2 |
| | -3 | | | -1 | | | +5 | |
| (-3) | | (+5) | (-2) | | (+3) | (+1) | | (-2) |

| | | | | | | | | |
|---|---|---|---|---|---|---|---|---|
| | 02 | | | 03 | | | 06 | |
| | 3 8 | | | 4 8 | | | 2 4 | x2 |
| | +5 | | | +4 | | | +2 | |
| (+2) | | (+1) | (+2) | | (-7) | (-2) | | (+2) |

| | | | |
|---|---|---|---|
| 05 | 07 | 06 | |
| 5 9 | 6 1 | 0 6 | x3 |
| +4 | -5 | +6 | |

| B/F | W/B – 1 | W/B – 2 | W/B – 3 |
|---|---|---|---|
| 27-3-‘22 | 02 | 01 | 05 |
| 188 | 0 2 | 1 0 | 5 0 |
| 08 | +2 | -1 | -5 |
| | (+1) (-1) | (+0) (+8) | (-2) (+9) |
| 29-3-‘22 | 02 | 09 | 03 |
| 191 | 1 1 | 1 8 | 3 9 |
| 02 | +0 | +7 | +6 |
| | (-1) (+2) | (-1) (-1) | (-1) (-8) |
| 31-3-‘22 | 03 | 07 | 03 |
| 110 | 0 3 | 0 7 | 2 1 |
| 02 | +3 | +7 | -1 |
| | (+0) (+3) | (+2) (+1) | (+2) (+6) |
| 3-4-‘22 | 06 | 01 | 02 |
| 216 | 0 6 | 2 8 | 4 7 |
| 09 | +6 | +6 | +3 |
| | (-0) (-4) | (+1) (-6) | (-1) (+2) |
| 5-4-‘22 | 02 | 05 | 03 |
| 194 | 0 2 | 3 2 | 3 9 |
| 05 | +2 | -1 | +6 |
| | (+0) (+4) | (+1) (+0) | (+1) (-4) |

| 7-4-'22 | 06 | 06 | 09 |
|---|---|---|---|
| 222 | 0 6 | 4 2 | 4 5 |
| 06 | +6 | -2 | +1 |
| | (+0) (+0) | (-3) (+4) | (-1) (-4) |

| 10-4-'22 | 06 | 07 | 04 |
|---|---|---|---|
| 199 | 0 6 | 1 6 | 3 1 |
| 01 | +6 | +5 | -2 |
| | (-0) (-1) | (-1) (+1) | (-1) (+3) |

| 12-4-'22 | 05 | 07 | 06 |
|---|---|---|---|
| 105 | 0 5 | 0 7 | 2 4 |
| 06 | +5 | +7 | +2 |
| | (+1) (-1) | (+1) (-1) | (+2) (-3) |

| 14-4-'22 | 05 | 07 | 05 |
|---|---|---|---|
| 228 | 1 4 | 1 6 | 4 1 |
| 03 | +3 | +5 | -3 |

| W/B – 4 | W/B – 5 | R/B | |
|---|---|---|---|
| 05 | 07 | 06 | |
| 5 9 | 6 1 | 0 6 | x3 |
| +4 | -5 | +6 | |
| (-0) (-1) | (+0) (+1) | (-0) (-3) | |
| 04 | 08 | 03 | |
| 5 8 | 6 2 | 0 3 | x2 |
| +3 | -4 | +3 | |
| (-2) (-7) | (-3) (+5) | (+1) (-2) | |

04 01 02

3 1 3 7 1 1 x3

-2 +4 +0

(+2) (+7) (+2) (+2) (+0) (+7)

04 05 09

5 8 5 9 1 8 x2

+3 +4 +7

(-1) (-2) (+1) (+0) (-1) (-2)

01 06 06

4 6 6 9 0 6 x2

+2 +3 +6

(+0) (+1) (-0) (-5) (+1) (+2)

02 01 09

4 7 6 4 1 8 x3

+3 -2 +7

(+2) (-5) (+0) (+2) (+0) (+0)

08 03 09

6 2 6 6 1 8 x2

-4 +0 +7

(-3) (-1) (-3) (-2) (-1) (-4)

04 07 04

3 1 3 4 0 4 x2

-2 +1 +4

(+3) (+2) (+3) (+4) (+2) (+2)

| | | | |
|---|---|---|---|
| 09 | 05 | 08 | |
| 6 3 | 6 8 | 2 6 | x2 |
| -3 | +2 | +4 | |

| B/F | | W/B – 1 | | | W/B – 2 | | | W/B – 3 | |
|---|---|---|---|---|---|---|---|---|---|
| 14-4-'22 | | 05 | | | 07 | | | 05 | |
| 228 | | 1 4 | | | 1 6 | | | 4 1 | |
| 03 | | +3 | | | +5 | | | -3 | |
| | (+0) | | (+1) | (+1) | | (-5) | (-1) | | (+1) |
| 17-4-'22 | | 06 | | | 03 | | | 05 | |
| 221 | | 1 5 | | | 2 1 | | | 3 2 | |
| 05 | | +4 | | | -1 | | | -1 | |
| | (-1) | | (+3) | (+1) | | (+2) | (+2) | | (+3) |
| 19-4-'22 | | 08 | | | 06 | | | 01 | |
| 235 | | 0 8 | | | 3 3 | | | 5 5 | |
| 01 | | +8 | | | +0 | | | +0 | |
| | (+2) | | (-8) | (-0) | | (-3) | (-1) | | (-0) |
| 21-4-'22 | | 02 | | | 03 | | | 09 | |
| 220 | | 2 0 | | | 3 0 | | | 4 5 | |
| 04 | | -2 | | | -3 | | | +1 | |
| | (-1) | | (-0) | (+0) | | (+9) | (+0) | | (+2) |
| 24-4-'22 | | 01 | | | 03 | | | 02 | |
| 209 | | 1 0 | | | 3 9 | | | 4 7 | |
| 02 | | -1 | | | +6 | | | +3 | |
| | (+0) | | (+2) | (-2) | | (-1) | (-2) | | (-7) |

| 26-4-'22 | 03 | 09 | 02 |
|---|---|---|---|
| 160 | 1 2 | 1 8 | 2 0 |
| 07 | +1 | +7 | -2 |
| | (-0) (-1) | (+2) (-2) | (+4) (+1) |

| 28-4-'22 | 02 | 09 | 07 |
|---|---|---|---|
| 242 | 1 1 | 3 6 | 6 1 |
| 08 | +0 | +3 | -5 |
| | (+0) (+3) | (-1) (-5) | (-3) (+6) |

| 1-5-'22 | 05 | 03 | 01 |
|---|---|---|---|
| 180 | 1 4 | 2 1 | 3 7 |
| 09 | +3 | -1 | +4 |
| | (+0) (+4) | (+0) (+6) | (-0) (-4) |

| 3-5-'22 | 09 | 09 | 06 |
|---|---|---|---|
| 169 | 1 8 | 2 7 | 3 3 |
| 07 | +7 | +5 | +0 |

| W/B – 4 | W/B – 5 | R/B | |
|---|---|---|---|
| 09 | 05 | 08 | |
| 6 3 | 6 8 | 2 6 | x2 |
| -3 | +2 | +4 | |
| (-0) (-1) | (-0) (-3) | (+0) (+0) | |
| 08 | 02 | 08 | |
| 6 2 | 6 5 | 2 6 | x5 |
| -4 | -1 | +4 | |
| (-1) (+7) | (-0) (-3) | (-1) (+2) | |

| | | | |
|---|---|---|---|
| 05 | 08 | 09 | |
| 5 9 | 6 2 | 1 8 | x2 |
| +4 | -4 | +7 | |
| (-0) (-4) | (-1) (+4) | (-0) (-4) | |

| | | | |
|---|---|---|---|
| 01 | 02 | 05 | |
| 5 5 | 5 6 | 1 4 | x2 |
| +0 | +1 | +3 | |
| (-1) (+4) | (+0) (+0) | (-1) (+4) | |

| | | | |
|---|---|---|---|
| 04 | 02 | 08 | |
| 4 9 | 5 6 | 0 8 | x3 |
| +5 | +1 | +8 | |
| (-1) (-0) | (+1) (-5) | (+1) (-8) | |

| | | | |
|---|---|---|---|
| 03 | 07 | 01 | |
| 3 9 | 6 1 | 1 0 | X2 |
| +6 | -5 | -1 | |
| (+3) (-7) | (+0) (+7) | (-1) (+4) | |

| | | | |
|---|---|---|---|
| 08 | 05 | 04 | |
| 6 2 | 6 8 | 0 4 | x2 |
| -4 | +2 | +4 | |
| (-2) (+2) | (-0) (-5) | (-0) (-3) | |

| | | | |
|---|---|---|---|
| 08 | 09 | 01 | |
| 4 4 | 6 3 | 0 1 | x3 |
| +0 | -3 | +1 | |
| (-1) (+5) | (-2) (+1) | (+0) (+7) | |

| 03 | 08 | 08 | |
|---|---|---|---|
| 3 9 | 4 4 | 0 8 | x5 |
| +6 | +0 | +8 | |

| B/F | W/B – 1 | W/B – 2 | W/B – 3 |
|---|---|---|---|
| 3-5-‘22 | 09 | 09 | 06 |
| 169 | 1 8 | 2 7 | 3 3 |
| 07 | +7 | +5 | +0 |
| | (+2) (-1) | (+1) (+2) | (+2) (+2) |
| 5-5-‘22 | 01 | 03 | 01 |
| 286 | 3 7 | 3 9 | 5 5 |
| 07 | +4 | +6 | +0 |
| | (-3) (-3) | (-3) (-4) | (-5) (+1) |
| 8-5-‘22 | 04 | 05 | 06 |
| 120 | 0 4 | 0 5 | 0 6 |
| 03 | +4 | +5 | +6 |
| | (+1) (+4) | (+3) (-5) | (+3) (-1) |
| 10-5-‘22 | 09 | 03 | 08 |
| 196 | 1 8 | 3 0 | 3 5 |
| 07 | +7 | -3 | +2 |
| | (-1) (-3) | (-3) (+7) | (+3) (-4) |
| 12-5-‘22 | 05 | 07 | 07 |
| 223 | 0 5 | 0 7 | 6 1 |
| 07 | +5 | +7 | -5 |
| | (+0) (+1) | (+4) (-7) | (-2) (-0) |

| 15-5-'22 | 06 | 04 | 05 |
|---|---|---|---|
| 193 | 0 6 | 4 0 | 4 1 |
| 04 | +6 | -4 | -3 |

(+0) (+1) (-3) (+5) (-2) (+1)

| 17-5-'22 | 07 | 06 | 04 |
|---|---|---|---|
| 157 | 0 7 | 1 5 | 2 2 |
| 04 | +7 | +4 | +0 |

(+4) (-7) (+3) (-4) (+3) (+6)

| 19-5-'22 | 04 | 05 | 04 |
|---|---|---|---|
| 285 | 4 0 | 4 1 | 5 8 |
| 06 | -4 | -3 | +3 |

(-3) (+4) (-3) (+4) (-3) (-3)

| 22-5-'22 | 05 | 06 | 07 |
|---|---|---|---|
| 175 | 1 4 | 1 5 | 2 5 |
| 04 | +3 | +4 | +3 |

| W/B – 4 | W/B – 5 | R/B | |
|---|---|---|---|
| 03 | 08 | 08 | |
| 3 9 | 4 4 | 0 8 | x5 |
| +6 | +0 | +8 | |

(+3) (-6) (+2) (+5) (+2) (-5)

| 09 | 06 | 05 | |
|---|---|---|---|
| 6 3 | 6 9 | 2 3 | x2 |
| -3 | +3 | +1 | |

(-4) (+5) (-0) (-2) (-1) (-3)

01
2 8
+6
(+3) (-6)

04
6 7
+1
(-1) (-1)

01
1 0 x2
-1
(-1) (+5)

07
5 2
-3
(+1) (+1)

02
5 6
+1
(+1) (+3)

05
0 5 x2
+5
(+1) (+3)

09
6 3
-3
(-2) (+2)

06
6 9
+3
(-1) (-7)

09
1 8 x2
+7
(-1) (+1)

09
4 5
+1
(-1) (+1)

07
5 2
-3
(+1) (+2)

09
0 9 x3
+9
(+1) (-6)

09
3 6
+3
(+3) (-2)

01
6 4
-2
(+0) (+1)

04
1 3 x2
+2
(+0) (+4)

01
6 4
-2
(-1) (-2)

02
6 5
-1
(-1) (+3)

08
1 7 x3
+6
(-0) (-6)

| 07 | 04 | 02 | |
|---|---|---|---|
| 5 2 | 5 8 | 1 1 | x2 |
| -3 | +3 | +0 | |

| B/F | | W/B – 1 | | | W/B – 2 | | | W/B – 3 | |
|---|---|---|---|---|---|---|---|---|---|
| 22-5-'22 | | 05 | | | 06 | | | 07 | |
| 175 | | 1 4 | | | 1 5 | | | 2 5 | |
| 04 | | +3 | | | +4 | | | +3 | |
| | (-1) | | (-3) | (+2) | | (-2) | (+1) | | (+2) |
| 24-5-'22 | | 01 | | | 06 | | | 01 | |
| 178 | | 0 1 | | | 3 3 | | | 3 7 | |
| 07 | | +1 | | | +0 | | | +4 | |
| | (+1) | | (+8) | (-1) | | (+5) | (+0) | | (+2) |
| 26-5-'22 | | 01 | | | 01 | | | 03 | |
| 202 | | 1 9 | | | 2 8 | | | 3 9 | |
| 04 | | +8 | | | +6 | | | +6 | |
| | (-1) | | (-7) | (+1) | | (+1) | (+2) | | (-9) |
| 29-5-'22 | | 02 | | | 03 | | | 05 | |
| 233 | | 0 2 | | | 3 9 | | | 5 0 | |
| 08 | | +2 | | | +6 | | | -5 | |
| | (+2) | | (+5) | (-1) | | (-1) | (+0) | | (+1) |
| 31-5-'22 | | 09 | | | 01 | | | 06 | |
| 265 | | 2 7 | | | 2 8 | | | 5 1 | |
| 04 | | +5 | | | +6 | | | -4 | |
| | (-1) | | (-6) | (+2) | | (-7) | (+0) | | (+5) |

| 2-6-'22 | 02 | 05 | 02 |
|---|---|---|---|
| 230 | 1 1 | 4 1 | 5 6 |
| 05 | +0 | -3 | +1 |

(+0) (+3) (-3) (+5) (-2) (-0)

| 5-6-'22 | 05 | 07 | 09 |
|---|---|---|---|
| 194 | 1 4 | 1 6 | 3 6 |
| 05 | +3 | +5 | +3 |

(-1) (-2) (-0) (-6) (-0) (-1)

| 7-6-'22 | 02 | 01 | 08 |
|---|---|---|---|
| 141 | 0 2 | 1 0 | 3 5 |
| 06 | +2 | -1 | +2 |

(+2) (+0) (+2) (+9) (+1) (-2)

| 9-6-'22 | 04 | 03 | 07 |
|---|---|---|---|
| 237 | 2 2 | 3 9 | 4 3 |
| 03 | +0 | +6 | -1 |

| W/B – 4 | W/B – 5 | R/B | |
|---|---|---|---|
| 07 | 04 | 02 | |
| 5 2 | 5 8 | 1 1 | x2 |
| -3 | +3 | +0 | |

(-2) (+7) (-1) (-6) (+1) (+5)

| 03 | 06 | 08 | |
|---|---|---|---|
| 3 9 | 4 2 | 2 6 | x2 |
| +6 | -2 | +4 | |

(+1) (-7) (+1) (+5) (-1) (+1)

| 06 | 03 | 08 | |
|---|---|---|---|
| 4 2 | 5 7 | 1 7 | x3 |
| -2 | +2 | +6 | |
| (+2) (-1) | (+1) (-1) | (-0) (-2) | |

| 07 | 03 | 06 | |
|---|---|---|---|
| 6 1 | 6 6 | 1 5 | x2 |
| -5 | +0 | +4 | |
| (+0) (+7) | (+0) (+3) | (+1) (-3) | |

| 05 | 06 | 04 | |
|---|---|---|---|
| 6 8 | 6 9 | 2 2 | x2 |
| +2 | +3 | +0 | |
| (-1) (-1) | (-0) (-6) | (-2) (-0) | |

| 03 | 09 | 02 | |
|---|---|---|---|
| 5 7 | 6 3 | 0 2 | x2 |
| +2 | -3 | +2 | |
| (-0) (-5) | (-0) (-3) | (+1) (+4) | |

| 07 | 06 | 07 | |
|---|---|---|---|
| 5 2 | 6 0 | 1 6 | x3 |
| -3 | -6 | +5 | |
| (-1) (+2) | (-2) (+6) | (-1) (-2) | |

| 08 | 01 | 04 | |
|---|---|---|---|
| 4 4 | 4 6 | 0 4 | x2 |
| +0 | +2 | +4 | |
| (+2) (-2) | (+2) (-2) | (+0) (+3) | |

| 08 | 01 | 07 | |
|---|---|---|---|
| 6 2 | 6 4 | 0 7 | x4 |
| -4 | -2 | +7 | |

| B/F | W/B – 1 | W/B – 2 | W/B – 3 |
|---|---|---|---|
| 9-6-‘22 | 04 | 03 | 07 |
| 237 | 2 2 | 3 9 | 4 3 |
| 03 | +0 | +6 | -1 |
| | (-1) (+6) | (-1) (-9) | (-2) (+3) |
| 12-6-‘22 | 09 | 02 | 08 |
| 191 | 1 8 | 2 0 | 2 6 |
| 02 | +7 | -2 | +4 |
| | (-1) (-6) | (+0) (+7) | (+2) (-4) |
| 14-6-‘22 | 02 | 09 | 06 |
| 191 | 0 2 | 2 7 | 4 2 |
| 02 | +2 | +5 | -2 |
| | (+1) (+7) | (+0) (+1) | (-0) (-1) |
| 16-6-‘22 | 01 | 01 | 05 |
| 188 | 1 9 | 2 8 | 4 1 |
| 08 | +8 | +6 | -3 |
| | (-0) (-9) | (-1) (+1) | (-0) (-1) |
| 19-6-‘22 | 01 | 01 | 04 |
| 197 | 1 0 | 1 9 | 4 0 |
| 08 | -1 | +8 | -4 |
| | (-1) (+3) | (+3) (-5) | (+2) (+1) |

| 21-6-'22 | 03 | 08 | 07 |
|---|---|---|---|
| 253 | 0 3 | 4 4 | 6 1 |
| 01 | +3 | +0 | -5 |
| | (+0) (+3) | (-3) (-4) | (-3) (-0) |

| 23-6-'22 | 06 | 01 | 04 |
|---|---|---|---|
| 163 | 0 6 | 1 0 | 3 1 |
| 01 | +6 | -1 | -2 |
| | (+0) (+0) | (+0) (+2) | (-1) (-1) |

| 26-6-'22 | 06 | 03 | 02 |
|---|---|---|---|
| 101 | 0 6 | 1 2 | 2 0 |
| 02 | +6 | +1 | -2 |
| | (-0) (-5) | (+0) (+1) | (-1) (+8) |

| 28-6-'22 | 02 | 04 | 09 |
|---|---|---|---|
| 125 | 1 1 | 1 3 | 1 8 |
| 08 | +0 | +2 | +7 |

| W/B – 4 | W/B – 5 | R/B | |
|---|---|---|---|
| 08 | 01 | 07 | |
| 6 2 | 6 4 | 0 7 | x4 |
| -4 | -2 | +7 | |
| (-1) (+1) | (+0) (+5) | (-0) (-2) | |
| 08 | 06 | 05 | |
| 5 3 | 6 9 | 0 5 | x2 |
| -2 | +3 | +5 | |
| (-1) (+1) | (-1) (-8) | (+2) (+0) | |

08
4 4
+0
(-0) (-2)

06
5 1
-4
(+0) (+0)

07
2 5 x2
+3
(-2) (+2)

06
4 2
-2
(+0) (+3)

06
5 1
-4
(+0) (+7)

07
0 7 x2
+7
(+2) (-2)

09
4 5
+1
(+2) (-2)

04
5 8
+3
(+1) (+1)

07
2 5 x2
+3
(-1) (-2)

09
6 3
-3
(-2) (+5)

06
6 9
+3
(-1) (-3)

04
1 3 x2
+2
(-0) (-1)

03
4 8
+4
(-2) (-1)

02
5 6
+1
(-2) (-4)

03
1 2 x3
+1
(-1) (+2)

09
2 7
+5
(+1) (-7)

05
3 2
-1
(+0) (+5)

04
0 4 x3
+4
(+1) (+2)

| 03 | 01 | 07 | |
|---|---|---|---|
| 3 0 | 3 7 | 1 6 | x3 |
| -3 | +4 | +5 | |

| B/F | W/B – 1 | W/B – 2 | W/B – 3 |
|---|---|---|---|
| 28-6-'22 | 02 | 04 | 09 |
| 125 | 1 1 | 1 3 | 1 8 |
| 08 | +0 | +2 | +7 |
| | (-1) (+7) | (+3) (-3) | (+3) (+1) |
| 30-6-'22 | 08 | 04 | 04 |
| 232 | 0 8 | 4 0 | 4 9 |
| 07 | +8 | -4 | +5 |
| | (+0) (+1) | (-3) (-0) | (-1) (-2) |
| 3-7-'22 | 09 | 01 | 01 |
| 203 | 0 9 | 1 0 | 3 7 |
| 05 | +9 | -1 | +4 |
| | (+1) (-4) | (+0) (+6) | (-1) (-3) |
| 5-7-'22 | 06 | 07 | 06 |
| 146 | 1 5 | 1 6 | 2 4 |
| 02 | +4 | +5 | +2 |
| | (+2) (-3) | (+2) (+0) | (+2) (+5) |
| 7-7-'22 | 05 | 09 | 04 |
| 261 | 3 2 | 3 6 | 4 9 |
| 09 | -1 | +3 | +5 |
| | (-2) (+2) | (-1) (-4) | (-0) (-7) |

| 10-7-'22 | 05 | 04 | 06 |
|---|---|---|---|
| 200 | 1 4 | 2 2 | 4 2 |
| 02 | +3 | +0 | -2 |

(-1) (-0) (+0) (+4) (-1) (+2)

| 12-7-'22 | 04 | 08 | 07 |
|---|---|---|---|
| 162 | 0 4 | 2 6 | 3 4 |
| 09 | +4 | +4 | +1 |

(+2) (-2) (-0) (-3) (+0) (+2)

| 14-7-'22 | 04 | 05 | 09 |
|---|---|---|---|
| 193 | 2 2 | 2 3 | 3 6 |
| 04 | +0 | +1 | +3 |

(-2) (+1) (-1) (+5) (-1) (-3)

| 17-7-'22 | 03 | 09 | 05 |
|---|---|---|---|
| 154 | 0 3 | 1 8 | 2 3 |
| 01 | +3 | +7 | +1 |

| W/B – 4 | W/B – 5 | R/B | |
|---|---|---|---|
| 03 | 01 | 07 | |
| 3 0 | 3 7 | 1 6 | x3 |
| -3 | +4 | +5 | |

(+2) (+8) (+3) (-4) (-0) (-2)

| 04 | 09 | 05 | |
|---|---|---|---|
| 5 8 | 6 3 | 1 4 | x3 |
| +3 | -3 | +3 | |

(+0) (+1) (-0) (-1) (+1) (+2)

| | | |
|---|---|---|
| 05<br>5 9<br>+4<br>(-2) (-8) | 08<br>6 2<br>-4<br>(-1) (+4) | 08<br>2 6 x3<br>+4<br>(-2) (-2) |
| 04<br>3 1<br>-2<br>(+3) (+1) | 02<br>5 6<br>+1<br>(+1) (+3) | 04<br>0 4 x2<br>+4<br>(+1) (-1) |
| 08<br>6 2<br>-4<br>(-2) (+4) | 06<br>6 9<br>+3<br>(-1) (-7) | 04<br>1 3 x2<br>+2<br>(+1) (+1) |
| 01<br>4 6<br>+2<br>(-1) (+1) | 07<br>5 2<br>-3<br>(+0) (+0) | 06<br>2 4 x3<br>+2<br>(-2) (+5) |
| 01<br>3 7<br>+4<br>(+1) (+0) | 07<br>5 2<br>-3<br>(+1) (+1) | 09<br>0 9 x2<br>+9<br>(-0) (-7) |
| 02<br>4 7<br>+3<br>(-1) (-5) | 09<br>6 3<br>-3<br>(-1) (+4) | 02<br>0 2 x2<br>+2<br>(+2) (-1) |

| 05 | 03 | 03 | |
|---|---|---|---|
| 3 2 | 5 7 | 2 1 | x2 |
| -1 | +2 | -1 | |

| B/F | W/B – 1 | W/B – 2 | W/B – 3 |
|---|---|---|---|
| 17-7-'22 | 03 | 09 | 05 |
| 154 | 0 3 | 1 8 | 2 3 |
| 01 | +3 | +7 | +1 |
| | (+1) (+1) | (+2) (-4) | (+1) (+3) |
| 19-7-'22 | 05 | 07 | 09 |
| 197 | 1 4 | 3 4 | 3 6 |
| 08 | +3 | +1 | +3 |
| | (-0) (-4) | (-1) (-4) | (-1) (-3) |
| 21-7-'22 | 01 | 02 | 05 |
| 189 | 1 0 | 2 0 | 2 3 |
| 09 | -1 | -2 | +1 |
| | (+2) (+9) | (+2) (+1) | (+3) (+1) |
| 24-7-'22 | 03 | 05 | 09 |
| 267 | 3 9 | 4 1 | 5 4 |
| 06 | +6 | -3 | -1 |
| | (-1) (-4) | (-1) (+6) | (-2) (+4) |
| 26-7-'22 | 07 | 01 | 02 |
| 209 | 2 5 | 3 7 | 3 8 |
| 02 | +3 | +4 | +5 |
| | (-2) (-4) | (-1) (-2) | (+1) (-4) |

| 28-7-'22 | 01 | 07 | 08 |
|---|---|---|---|
| 208 | 0 1 | 2 5 | 4 4 |
| 01 | +1 | +3 | +0 |
| | (+0) (+3) | (-1) (+2) | (+1) (+3) |

| 31-7-'22 | 04 | 08 | 03 |
|---|---|---|---|
| 216 | 0 4 | 1 7 | 5 7 |
| 09 | +4 | +6 | +2 |
| | (+1) (+1) | (+1) (-6) | (-2) (-6) |

| 2-8-'22 | 06 | 03 | 04 |
|---|---|---|---|
| 184 | 1 5 | 2 1 | 3 1 |
| 04 | +4 | -1 | -2 |
| | (-1) (+4) | (+0) (+0) | (+2) (+5) |

| 4-8-'22 | 09 | 03 | 02 |
|---|---|---|---|
| 220 | 0 9 | 2 1 | 5 6 |
| 04 | +9 | -1 | +1 |

| W/B – 4 | W/B – 5 | R/B | |
|---|---|---|---|
| 05 | 03 | 03 | |
| 3 2 | 5 7 | 2 1 | x2 |
| -1 | +2 | -1 | |
| (+2) (-2) | (+0) (+1) | (-2) (+4) | |
| 05 | 04 | 05 | |
| 5 0 | 5 8 | 0 5 | x3 |
| -5 | +3 | +5 | |
| (-1) (+9) | (+1) (-3) | (+2) (-3) | |

## American Powerball Lottery Results

04 02 04

4 9 | 6 5 | 2 2 x3

+5 -1 +0

(+1) (+0) (-0) (-3) (-1) (-0)

05 08 03

5 9 | 6 2 | 1 2 x3

+4 -4 +1

(-2) (-0) (+0) (+3) (-1) (+3)

03 02 05

3 9 | 6 5 | 0 5 x2

+6 -1 +5

(+2) (-4) (-1) (+2) (+2) (+1)

01 03 08

5 5 | 5 7 | 2 6 x2

+0 +2 +4

(+0) (+3) (+1) (+1) (-1) (-4)

04 05 03

5 8 | 6 8 | 1 2 x3

+3 +2 +1

(-2) (-2) (-0) (-3) (+0) (+4)

09 02 07

3 6 | 6 5 | 1 6 x3

+3 -1 +5

(+2) (+1) (+0) (+1) (-0) (-5)

| 03 | 03 | 02 | |
|---|---|---|---|
| 5 7 | 6 6 | 1 1 | x2 |
| +2 | +0 | +0 | |

# CONCLUSION

After reading my book thoroughly, you will see millions of odds reduced to less than a thousand.

Practice carefully, consistently, confidently and diligently every day for two to three months, as described in my book. Then, miraculously, you will see the odds of millions reduced to less than a hundred.

Even buying less than a hundred lottery tickets can cost more. So, form your own syndicate with your family members, friends and colleagues to keep your monthly lottery budget in line with your purchasing power.

Sharing the prize money with your loved ones will double your happiness and pave the way for your syndicate and you to win the lottery again and again.

Winning the lottery again and again will motivate you to share your prize money with those who need it for their basic amenities in daily life and increase your happiness. No doubt this happiness will give meaning to your human birth and great satisfaction to your soul in human service.

Don't wait to pick up my book.

After picking up my book, start your work immediately with a jackpot winning attitude.

If you have any doubts, please email me at the following email address.

tssn11091962@gmail.com

I will reply soon and clear your doubts.

Greetings.

B.Sc., Graduate, interested in logical solving of various types of fun puzzles. In particular, he is more interested in logically solving number puzzles.

He realized that when individuals are allowed to choose the numbers of their choice, the lottery becomes a great fun numbers game and naturally falls under the analysis of logical reasoning.

Also, he believed that upcoming lottery winning numbers could only be traced from previously drawn lottery numbers.

He logically designed a simple 'Distance and Difference' strategy with simple mathematics and basic arithmetic symbols to increase the chances of winning the lottery and find the upcoming winning numbers from previously drawn lottery numbers.

Based on his findings, he publishes this book as an excellent guide book for the benefit of lottery players.

His great belief is that our universe is always ready to respond and reward those who think and act on innovative ideas with a positive attitude towards anything.

www.ingramcontent.com/pod-product-compliance
Lightning Source LLC
LaVergne TN
LVHW021155160826
845679LV00024B/2132

* 9 7 9 8 8 9 4 4 6 6 8 3 5 *